Lecture Notes in Mathematics

A collection of informal reports and seminars
Edited by A. Dold, Heidelberg and B. Eckmann, Zürich

C. T. Taam

170

R.M. Dudley, J. Feldman,
B. Kostant, R.P. Langlands, E.M. Stein

Lectures in Modern Analysis and Applications III

Edited by C. T. Taam, George Washington University

Springer-Verlag
Berlin · Heidelberg · New York 1970

ISBN 3-540-05284-4 Springer-Verlag Berlin · Heidelberg · New York
ISBN 0-387-05284-4 Springer-Verlag New York · Heidelberg · Berlin

 Library of Congress Catalog Card Number 64-54683. Printed in Germany.

Offsetdruck: Julius Beltz, Weinheim/Bergstr.

PREFACE

This lecture series in Modern Analysis and Applications was sponsored by the Consortium of Universities (American University, Catholic University, Georgetown University, George Washington University, and Howard University) in Washington, D.C. and the University of Maryland, in conjunction with the U.S. Air Force Office of Scientific Research. These lectures were presented at the sponsoring universities over the period 1967-69 by mathematicians who have contributed much to the recent growth of analysis.

The series consisted of eight (8) sessions of three lectures each. Each session was devoted to an active basic area of contemporary analysis which is important in application or shows potential applications. Each lecture presented a survey and a critical review of certain aspects of that area, with emphasis on new result, open problem, and application.

The eight sessions of the series were devoted to the following basic areas of analysis:

1. Modern Methods and New Results in Complex Analysis
2. Banach Algebras and Applications
3. Topological Linear Spaces and Applications
4. Geometric and Qualitative Aspects of Analysis
5. Analysis and Representation Theory
6. Modern Analysis and New Physical Theories
7. Modern Harmonic Analysis and Applications
8. Integration in Function Spaces and Applications

This volume contains five lectures from the seventh and eighth sessions. Some of the manuscripts were proof-read by the editor after they were retyped for reproduction, the editor would be grateful to have the authors' understanding in this matter which was due to the pressures of time.

C. T. Taam

George Washington University

ORGANIZING COMMITTEE

George Butcher
Howard University

Avron Douglis
University of Maryland

John E. Lagnese
Georgetown University

Raymond W. Moller
Catholic University of America

Robert G. Pohrer
U.S. Air Force Office of Scientific Research

Steven H. Schot
American University

C. T. Taam, Chairman
George Washington University

Elmer West
Consortium of Universities

TABLE OF CONTENTS

MODERN HARMONIC ANALYSIS AND APPLICATIONS

Professor E. M. STEIN, Princeton University

Variations on the Littlewood-Paley Theme 1

The purpose of this lecture will be to discuss a new approach to the classical Littlewood-Paley theory, viewed in the framework of symmetric diffusion semi-groups. This leads to a generalization of parts of that theory and further applications in a variety of new situations.

Professor R. P. LANGLANDS, Yale University

Problems in the Theory of Automorphic Forms 18

Analogues, for an automorphic form on a general reductive group, of the Hecke L-series will be introduced. Some questions in the theory of group representations to which they give rise will be discussed.

Professor BERTRAM KOSTANT, Massachusetts Institute of Technology

Quantization and Unitary Representations 87

We are concerned with the existence and irreducibility of unitary representations of Lie groups. Our general approach to unitary representations is by means of a quantization theory associated with certain symplectic manifolds. L. Auslander and I successfully applied this theory to obtain criteria for solvable Lie groups to be of type I and to determine all the irreducible unitary representations of such groups.

INTEGRATION IN FUNCTION SPACES AND APPLICATIONS

Professor R. M. DUDLEY, Massachusetts Institute of Technology

Random Linear Functionals: Some Recent Results 62

Given a linear mapping of a topological vector space S into random variables, when does it define a countably additive probability measure on the dual space S'? There are recent

results of Laurent Schwartz and co-workers. The Gaussian case leads to questions about measuring the sizes of compact, convex subsets of Hilbert space.

Professor JACOB FELDMAN, University of California, Berkeley

Absolute Continuity of Stochastic Processes........................ 71

Discussion of conditions for absolute continuity of measures induced by various classes of stochastic processes. Transformations of such processes. The notion of continuous product measure, and the study and classification of decomposable processes from this viewpoint.

Contents of Lectures in Modern Analysis and Applications I, II...... 209

VARIATIONS ON THE LITTLEWOOD-PALEY THEME

by E. M. Stein*

1. We shall use the phrase "Littlewood-Paley Theory" rather loosely, to describe a variety of related results in classical harmonic analysis whose extension to a more general setting will be the main subject of this lecture. We will begin by giving a review of the results for $\mathbb{R}^1$, then for $\mathbb{R}^n$, and finally sketch the elements of a recent general approach quite different from the earlier two. The details of this approach will be published soon; see [14].

2. The classical case of $\mathbb{R}^1$.

In its one-dimensional form the theory goes back to the 1930's, and may be said to contain the Hardy-littlewood maximal theorem, Hilbert transforms, the work of Littlewood-Paley, and capped in effect with the multiplier theorem of Marcinkiewicz. This theory may be described in terms of the Poisson integral which in R^1 is given by the family of transformations

$$f(x) \longrightarrow \frac{t}{\pi}\int_{-\infty}^{\infty} \frac{f(x-y)}{t^2+y^2}\,dy = u(x,t), \qquad t > 0.$$

As is well known, the behavior of harmonic functions $u(x,t)$ closely reflects the behavior of its boundary values. Now the main idea of the so-called "complex-method" is to pass to the analytic function $F(z)$, whose real part is u, then to exploit complex function theory to study F and thus f. An example of this far-reaching idea arises in the Hilbert transform, H,

$$(Hf)(x) = \lim_{\varepsilon \to 0} \frac{1}{\pi}\int_{|t| \geq \varepsilon} \frac{f(x-t)\,dt}{t}.$$

This transform gives in effect the passage from the real and imaginary parts of the boundary values of F.

---

*The author was partially supported by a grant of the Air Force Office of Scientific Research at Princeton University.

We list now some of the main results of the one-deminisional theory.

Theorem A.

Let $(Mf)(x) = \sup_{t>0} |u(x,t)|$. Then

$$\|Mf\|_p \leq A_p \|f\|_p , \qquad 1 < p \leq \infty$$

where $\|.\|_p$ denotes the usual L^p norm in R^1.

This theorem is an easy reformulation of the Hardy-Littlewood maximal theorem.

Theorem B.

$$\|H(f)\|_p \leq A_p \|f\|_p \qquad 1 < p < \infty.$$

This is M. Riesz's celebrated theorem about the Hilbert transform.

The Hilbert transform, of course, has a simple expression in L^2 in terms of the Fourier transform, namely $(Hf)\hat{\ }(x) = i \operatorname{sign} x \hat{f}(x)$. If now S_R denotes the "partial sum" operator given by

$$(S_R(f)\hat{\ }(x) = \chi_R(x) \hat{f}(x)$$

where χ is the characteristic function of $|x| \leq R$, then an immediate corollary of Theorem B is

Corollary C $\qquad \|S_R(f)\|_p \leq A_p \|f\|_p , \qquad 1 < p < \infty$

with A_p independent of R.

The next series of results deal with what is commonly known as the Littlewood-Paley theory, and go back to their papers [7]; see also [17], Chapter 14.

One of their main ideas is to introduce a certain quadratic auxiliary operator, whose application turns out to be a very powerful tool. A typical example is their g-function, defined as

$$g(f)(x) = \left(\int_0^\infty |\nabla u(x,t)|^2 t\,dt \right)^{1/2}$$

where $$|\nabla u|^2 = \left|\frac{\partial u}{\partial x}\right|^2 + \left|\frac{\partial u}{\partial t}\right|^2$$

with $u(x,t)$ the Poisson integral of f. A main technical result is contained in the following inequality

Theorem D.

$$B_p \|f\| \leq \|g(f)\|_p \leq A_p \|f\|, \qquad 1 < p < \infty$$

There are many variants of this that are possible. One is related to another auxiliary operator, the "area integral" of Lusin, defined by

$$S(f)(x_o) = \left(\iint_{\Gamma(x_o)} |\nabla u|^2 dx\,dt \right)^{1/2}$$

where $\Gamma(x_o)$ is the angle $\{|x - x_o| < \alpha t,\ t > 0,\}$, for some fixed α. It is not hard to see that

$$g(f)(x) \leq \text{constant } S(f)(x).$$

Moreover, inequalities analogous to that for g hold for S. A somewhat different variant which is also interesting, arises if we consider the Hardy classes H^p, $p > 0$. We are now dealing with functions that are boundary values at holomorphic functions (we call these F to indicate the change in situation). Then one can also prove

Corollary E. $\|g(F)\|_p \leq A_p \|F\|_p \quad 0 < p < \infty$, and

similarly $\|S(F)\|_p \leq A_p \|F\|_p \quad , 0 < p < \infty$.

Notice that at this stage the converse inequality (for $p \leq 1$) is not claimed (and was then not known).

Now Littlewood and Paley made use of all the above results and in particular theorems C and D to give their dyadic decompositional Fourier integrals which can be stated as follows.

Let $\{I_j\}$ denote the collection of intervals $(2^k, 2^{k+1})_{k=-\infty}^{\infty}$ and $(-2^{k+1}, -2^k)_{k=-\infty}^{\infty}$. Let $S_{I_j}(f)$ be defined by $(S_{I_j} f)^{\wedge} = \chi_{I_j}(x) \hat{f}(x)$, where χ_{I_j} is the characteristic function of I_j.

Theorem F.

$$B_p \|f\|_p \leq \|(\sum_j |S_{I_j} f|^2)^{1/2}\|_p \leq A_p \|f\|_p, \quad 1 < p < \infty.$$

The thrust of this theorem is that it shows that the different dyadic blocks $S_{I_j}(f)$ of the Fourier expansion f behave as if they were (very roughly speaking) independent, or, somewhat closer to the truth, as if at least they were strongly orthogonal.

Theorem F was exploited by Marcinkiewicz who proved a general multiplier theorem which essentially contains theorems B, C, D, F, and which in addition has a definite interest of its own. Let $m(x)$ be defined on $\mathbb{R}^1$ and assume that m is of bounded variation in every finite interval excluding the origin. Assume in addition that

$$(1) \quad |m(x)| \leq A \quad \text{and} \quad \int_{R \leq |x| \leq 2R} |dm(x)| \leq A, \quad \text{all } R > 0.$$

We define the multiplier operator T_m, by

$$(T_m f)^{\wedge}(x) = m(x) \hat{f}(x), \qquad \text{if } f \in L^2 \cap L^p.$$

Theorem G.

$$\|T_m(f)\|_p \leq A \|f\|_p \qquad 1 < p < \infty .$$

A brief historical remark is in order. The above theorems were developed in the 1930's. This development was carried out in the context of the circle (i.e. Fourier series) instead of the real line (Fourier integrals), but the passage from the first to the latter does not involve essential difficulties; or if one wishes, one can carry out the argument in a completely parallel fashion to that of the circle. For detailed proofs in the periodic case, see Zygmund's book [17], in particular Chapters 14 and 15.

Before we leave the classical case of R^1 we should like to mention that there have been recently some results in this area, as curious as this may seem. The first result is the converse inequality to Corollary E, valid for all $p > 0$, but whose principal interest is the case $p = 1$, namely the following theorem of Calderón [2]:

Theorem H.

Supoose $F \in H^p$, $0 < p$, then

$$\|F\|_p \leq A_p \|S(F)\|_p .$$

Another result is in part a consequence of this, and shows that an important portion of the above theory can be extended to the Hardy clases H^p, where $p \leq 1$: see [12]. I mention one result of this type, the analogue of the Marcinkiewicz multiplier theorem (Theorem G) to the H^p setting.

Here the definition of m only for $x > 0$ matters. We assume that $m(x)$ has k derivaties which are square integrable in every finite interval excluding the origin and,

$$(2) \qquad |m(x)| \leq A \qquad \int_{R \leq x \leq 2R} |m^{(\ell)}(x)|^2 dx \leq AR^{-2\ell+1}, \quad \ell=0,1,\ldots,k.$$

Theorem I.

$$\|T_m(F)\|_p \leq A\|F\|_p, \qquad \text{if } p > \frac{1}{k}.$$

An interesting example arises as follows:

Example J.

Let $m(x) = x^{i\gamma}$, for γ real

Then $\|T_m(F)\|_p \leq A_{p,\gamma} \|F\|_p$, all $p > 0$.

3. The "semi-classical" case $\mathbb{R}^n$.

The next stage in the progress of this type of analysis was in the context of $\mathbb{R}^n$, and reached its high point in about the 1950's. It was the primacy of complex function theory give way to real-variable methods.

Important aspects of these real-methods had been anticipated earlier by Besicovitch and Marcinkiewicz; the former in his real-variable approach to the Hilbert transform, and the latter in the same area and in his development of the interpolation theorem which bears his name and which has since become a very fundamental tool in the entire subject.

Together with this should be mentioned the n-dimensional generalization of the maximal theorem of Hardy and Littlewood as given by Wiener [16], and the later lemma of Calderón and Zygmund in [3]. This lemma states that given on f integrable on $\mathbb{R}^n$, and any $\alpha > 0$, we can split f, $f = f_1+f_2$, so that: $|f_1| \lesssim \alpha$; f_2 is supported on a disjoint union of cubes with the property that the mean-value of $|f_2|$ on each of these cubes is approximately α.

We now briefly list the results that can be obtained analogous to R^1. First, the Poisson integral can be defined by $u(x,t) = \int_{R^n} f(x-y)P_t(y)\,dy$

where $P_t(y) = c_n \frac{t}{(t^2+|y|^2)^{\frac{n+1}{2}}}$, with $c_n = \frac{\Gamma(\frac{n+1}{2})}{\pi^{(n+1)}}$

Write $Mf(x) = \sup_{t>0} |u(x,t)|$. Then

Theorem A_n.

$$\|Mf\|_p \leq A_p \|f\|_p, \qquad \text{if} \quad 1 < p \leq \infty.$$

This is a consequence of Wiener's n-dimensional maximal theorem.

The n-dimensional form of the Hilbert transform goes back to Giraud, Mihlin, and Calderón and Zygmund. A special case can be formulated as follows: Let Ω be a function which is homogeneous of degree 0 in $\mathbb{R}^n$, which satisfies a certain minimal smoothness, and whose mean-value on the unit ball centered at the origin vanishes.

Then if $f \in L^p(\mathbb{R}^n)$, $1 < p < \infty$,

$$(Hf)(x) = \lim_{\varepsilon \to 0} \int_{|y| \geq \varepsilon} \frac{\Omega(y)}{|y|^n} f(x-y)\, dy$$

exists in L^p as we have

Theorem B_n.

$$\|H(f)\|_p \leq A_p \|f\|_p, \quad \underline{\text{if}} \quad 1 < p < \infty.$$

A further special case of this arises when $\Omega(x) = c_n \frac{x_j}{|x|}$ $j = 1, \ldots, n$. Then $H(f) = R_j(f)$, where the R_j are the "Riesz transforms."

The Fourier transform realization of the Riesz transforms is given by $(R_j f)^{\wedge}(x) = i \frac{x_j}{|x|} \hat{f}(x)$. Notice that we have formally

(3) $$R_j = \frac{\partial}{\partial x_j} (-\Delta)^{-1/2}.$$

It was observed by the author that the techniques used for the proof of Theorem B_n concerning singular integrals could be applied also to the g-functions and their analogues. Let

$$g(f)(x) \left(\int_0^\infty |\nabla u(x,t)|^2 dt \right)^{1/2}, \quad \text{and}$$

$$S(f)(x) = \left(\int_{R^n} \int_0^\infty \chi(x,x',t) |\nabla u(x',t)|^2 t^{1-n} dt dx' \right)^{1/2}$$

with χ the charactertic function of the cone $\{|x-x'| < \alpha t, t>0\}$. Thenit was shown in [11] that the following held.

Theorem D_n.

$$B_p \|f\|_p \leq \|g(f)\|_p \leq A_p \|f\|_p, \quad 1 < p < \infty$$

with a similar result for S in place of g.

Later papers have simplified these ideas and synthesized the proofs of Theorems B_n and D_n. See in particular [1].

There is no really satisfactory version in R^n of Corollary C. The situation concerning this is as follows. For certain sets such as cubes, or more generally polyhedra, the parallel to Corollary C can be proved as a consequence of the one-dimensional result. What is needed is the solution of the multiplier problem concerning the characteristic function of a ball; more particularly it may be conjectured that there is an analogue of Theorem C in this case (with the ball taking the place of the interval $(-R,R)$), for $2n/(n+1) < p < 2n/(n-1)$.

For the same reason there is no fully satisfactory version of Theorems F and G to n-dimensions. However, as Marcinkiewicz earlier showed in [8], certain variants could be obtained from the one-dimensional theorems. Despite the limitations alluded to, the resulting multiplier theorem had very interesting implications and in effect essentially contained the result in Theorem B_n when the kernel $\frac{\Omega(x)}{|x|^n}$ is sufficiently smooth (away from the origin). The original presentation in the context of multiple Fourier series is in [8]; its strict analogue for $\mathbb{R}^n$ is given in the author's notes [13]. We state here another version, going back to Mihlin and in the sharper form derived by Hörmander. As before we denote by T_m the transformation given by

$$(T_m f)^{\wedge}(x) = m(x) f^{\wedge}(x), \qquad f \in L^2(\mathbb{R}^n) \cap L^p(\mathbb{R}^n).$$

We assume that m has its partial derivatives of total order not greater than k square integrable on compact subsets not containing the origin and

$$(4) \quad |m(x)| \leq A, \int_{R \leq |x| \leq 2R} \left| \frac{\partial^{\ell}}{\partial x^{\ell}} m(x) \right|^2 dx \leq AR^{-2|\ell|+n}, \quad 0 \leq |\ell| \leq k$$

where k is the smallest integer $> n/2$.

<u>Theorem G_n</u>. <u>Assume that</u> m <u>satisfies</u> (4) <u>above</u>. <u>Then</u>

$$\|T_m(f)\|_p \leq A_p \|f\|_p \qquad 1 < p < \infty.$$

We shall now discuss the analogues of Corollary E, and Theorems H, I together with example J in n-dimensions.

It was found by Stein and Weiss in [15] that it is possible to develop a theory of H^p spaces in the context of R^n which had some of the features of the classical H^p theory, and where the characteristic role of the Hilbert transform was replaced by the Riesz transforms (3). The crucial restriction that had to be made was that $p > \frac{n-1}{n}$, (when n=1, we would have $p > 0$). An important point was that in every dimension the case p=1 was covered. Using this theory Gasper [5] and Segovia [10] obtained

<u>Corollary E_n</u>. <u>Suppose</u> $\frac{n-1}{n} < p < \infty$ <u>and</u> $F \in H^p(\mathbb{R}^n)$.

<u>Then</u> $\|S(F)\|_p \leq A_p \|F\|_p$.

In his paper Segovia also obtained the analogous converse:

Theorem H_n.

Suppose $\frac{n-1}{n} < p < \infty$, and $F \in H^p(\mathbb{R}^n)$.

Then $\|F\|_p \leq A_p \|S(F)\|_p$.

The analogues of Theorem I and Example J were given by us in [12] at the same time as their one-dimensional versions.

Theorem I_n. *Suppose* m *satisfies the condition* (4) *for some* $k > n/2$.

Then $\|T_m(F)\|_p \leq A_p \|F\|_p$, *if* $p > n/k$, *and* $p < \frac{n-1}{n}$.

Example J_n. *Let* $m(x) = |x|^{i\gamma}$ *for* γ *real*. *Then*

$$\|T_m(F)\|_p \leq A_{p\gamma} \|F\|_p, \quad \text{all } p > \frac{n-1}{n}.$$

We should like to mention here a principle brought to light by us in [12] and [13]. It is crucial for the proofs of Theorem I_n, J_n; can be used to prove Theorem G_n (which incidentally is contained in Theorem I_n); and is a germinal idea in the abstract treatment sketched below. Namely if g_λ^* is one of the versions of the g-function (more inclusive than the S function described above) we have the basic inequality

(5) $$S(T_m f)(x) \leq A\, g_\lambda^*(f)(x)$$

where m satisfies the conditon (4), with appropriate λ. The inequality (5) and its many possible variants immediately reduce the study of multiplier operators T_m to appropriate inequalities for the functions g, or S, or g_λ^* etc.

However, as indicated before, various questions of interest in R^n remain open. Some interesting progress in this general area has recently been made by Fefferman [4].

4. The general case; an abstract approach, (see [14]).

The new approach we have in mind is essentially different from the complex and real-variable methods whose results are described in Sections 2 and 3. While this general approach treats only an essential core of the subject, it can nevertheless be applied to many new and interesting situations.

Our starting point is the family of Poisson integrals, which form a semi-group of operators. That is we assume that we are given a family of operators $\{T^t\}_{t\geq 0}$, defined simultaneously on $L^p(M)$, $1 \leq p \leq \infty$, (M is an arbitrary measure space), with the property that

$$T^{t_1} \cdot T^{t_2} = T^{t_1+t_2}, \quad T^0 = I, \quad t_1 \geq 0, t_2 \geq 0.$$

In addition to a standard measureability in t we make the following basic assumptions

(I) T^t are contractions on $L^p(M)$, i.e. $\|T^t f\|_p \leq \|f\|_p$, $1 \leq p \leq \infty$

(II) T^t are symmetric, i.e. each T^t is self-adjoint on $L^2(M)$.

(III) T^t are positivity preserving, i.e. $T^t f \geq 0$, if $f \geq 0$.

(IV) $T^t(1) = 1$.

We call a semi-group $\{T^t\}$ satisfying (I) to (IV) a symmetric diffusion semi-group. In some of our results we shall be able to dispense with assumptions (III) and (IV), but (I) and (II) will be essential.

The task we set ourselves is to develop, as far as is possible, the analogues of the Littlewood-Paley theory in the context of these semi-groups. The interest in this arises from the multiplicity of examples of symmetric diffusion semi-groups and the consequence this theory has for the eigenfunction expansion of their infinitesimal generators. Here are some examples of such semi-groups:

Example 1. The Poisson integral for $\mathbb{R}^n$, (and $\mathbb{R}^1$) used in Sections 3 and 2. This example can be written in the form $T^t = e^{tA}$, where the infinitesimal generator A is $-(-\Delta)^{1/2}$. The closely related semi-group of the heat equation arises if $A = -\Delta$, and is also of the symmetric-diffusion type.

Example 2. Δ is the Laplace-Beltrami operator on any compact smooth Riemannian manifold. Then one can construct semi-groups of this type where $T^t = e^{t\Delta}$, and also where $T^t = e^{-t(-\Delta)^{1/2}}$. This can also be done when Δ is the Laplace-Beltrami operator on various non-compact Riemannian manifolds, or with manifolds with boundary, once appropriate boundary conditions are imposed.

Example 3. Special cases of interest in Example 2 arise if we take Δ to be the Laplace-Beltrami operator on a compact Lie group with a bi-invariant Riemannian metric, or of any homogeneous space of that group; also if Δ is the Laplace-Beltrami operator of a left-invariant Riemannian metric on an arbitrary Lie group.

Example 4. Here M is an interval (α_1, α_2) (possibly infinite) in $\mathbb{R}^1$.

$A = a(x)\dfrac{d^2}{dx^2} + b(x)\dfrac{d}{dx} + c(x)$ with $a(x)$, $b(x)$, and $c(x)$ continuous, $a(x) > 0$, $c(x) \leq 0$. One can then find a measure $d\mu(x)$ with respect to which A is formally self-adjoint, and appropriate boundary conditions for A, so that $T^t = e^{tA}$ can be constructed (on $L^p(\alpha_1, \alpha_2, d\mu)$). This semi-group satisfies conditions (I), (II), and (III); if in addition $c(x) \equiv 0$, then (IV) is also satisfied.

Example 5. If A is such that the semi-group $T^t = e^{tA}$ satisfies any of the conditions (I) through (IV) then by the principle of subordination, the semi-group $e^{-t(-A)^\alpha}$, with $0 < \alpha < 1$, also satisfies the same conditions.

The main tools that are used in studying the symmetric diffusion semi-groups are three-fold.

(i) The spectral representation in $L^2(M)$ of T^t as $T^t = \int_0^\infty e^{-\lambda t} dE(\lambda)$, for an appropriate resolution of the identity $E(\lambda)$. This is the direct substitute for the Fourier transform in $\mathbb{R}^n$.

(ii) Connections of the semi-groups T^t and certain auxiliary martingales and ergodic theorems. It is curious to point out that the general martingale techniques that are needed were to a significant extent already anticipated by Paley [9] in his study of the Walsh-Paley series, where the primeval instance of a theory of Littlewood-Paley was developed! It is to be pointed out that the dyadic partial sums S_{2^k} of a Walsh-Paley series form a simple example of a martingale sequence.*

(iii) Convexity properties of analytic families of operators that allow one to obtain L^p results by using both (i) and (ii). The following analogues of the results of Sections 2 and 3 can then be obtained. (For details in what follows see [14].)

Theorem A_*.

Suppose $\{T^t\}_{t>0}$ satisfies conditions (I) and (II).

Let $(Mf)(x) = \sup_{t>0} |T^t f(x)|$. Then

$$\|Mf\|_p \leq A_p \|f\|_p, \quad 1 < p \leq \infty$$

Denote by $E_o(f) = \lim_{\lambda \to 0^+} E(\lambda)f$, where $E(\lambda)$ is the spectral resulution entering in $T^t = \int_o^\infty e^{-\lambda t} dE(\lambda)$.

Theorem D_*. Let $g_1(f)(x) = \left(\int_o^\infty t \left| \frac{\partial T^t f}{\partial t} \right|^2 dt \right)^{1/2}$.

If $1 < p < \infty$, then $B_p \|f\|_p \leq \|g_1(f)\|_p + \|E_o(f)\|_p \leq A_p \|f\|_p$.

There is also an analogue of the Marcinkiewicz multiplier theorem (Theorems G and G_n of Sections 2 and 3), but it now requires a certain analyticity of the multiplier. However, this is what is to be expected from a universal theorem of

---

*See the general formulation in Gundy [6], and an earlier result of Burkholder [0].

this type.

For any bounded measurable function $m(\lambda)$ on (o,∞) we define T_m by $T_m = \int_0^\infty m(\lambda)\,dE(\lambda)$.

__Theorem G_*.__ $\|T_m(f)\|_p \leq A_p\|f\|_p$ $1 < p < \infty$ _if_ $m(\lambda) = \lambda\int_0^\infty e^{-\lambda t}M(t)\,dt$, _where_ M _is bounded on_ (o,∞).

This contains a generalization of Example J_n.

__Example J_*.__ _The function_ $m(\lambda) = \lambda^{i\gamma}$, γ real, _satisfies the conditions of the above theorem._

5. A complementary approach (See [14]).

Besides the abstract theory just described another approach may be considered, which is less general but on the other hand more fruitful in various instances. It arises whenever, roughly speaking, the square of the infinitesimal generator of T^t is a second order elliptic operator, i.e. a "Laplacian". When A is this Laplacian (in the x variables) then the semi-group corresponds to the equation $Au(x,t) + \frac{\partial^2 u}{\partial t^2} = 0$. In this case the interplay between the x-derivatives and the t-derivatives are also of interest and it is possible to study the natural generalizations of the Hilbert transforms (i.e. the analogue of the Riesz transforms) in this setting. This situation occurs in the context of compact Lie groups (and any of their homogeneous spaces), and to some extent also in the noncompact Lie groups (in particular for the zonal functions of semi-simple Lie groups); also for Sturm-Liouville expansions.

This attack has its starting point the universality of the identity

$$\Delta u^p = p(p-1)\, u^{p-2} |\nabla u|^2$$

where $\Delta = A + \frac{\partial^2}{\partial t^2}$, $u > 0$, $\Delta u = 0$, with ∇ the associated gradient. Also the maximal theorem (Theorem A_* in Section 3) is needed here.

In this context it is possible to prove a stronger version of Theorem D_* dealing with the Littlewood-Paley g-functions namely, if

$$g(f)(x) = \left(\int_0^\infty t |\nabla u| dt \right)^{1/2}$$

then

$$\|g(f)\|_p \leq A_p \|f\|_p, \qquad 1 < p < \infty .$$

It will be best to describe the matters further in terms of three interesting instances; the case of $\mathbb{R}^n$, the case of a compact Lie group; and the case of a semi-simple Lie group.

Case of $\mathbb{R}^n$. Here we take A to be the usual Laplacian in $\mathbb{R}^n$. We define the Riesz transforms R_j by $R_j = \frac{\partial}{\partial x_j} (-A)^{-1/2}$ as in equation (3). We observe that $g_1(R_j f)(x) \leq g(f)(x)$. (This is the version of inequality (5) that is appropriate here.) The inequalities for the g-functions then prove

$$\|R_j(f)\|_p \leq A_p \|f\|_p \qquad 1 < p < \infty.$$

This model of proof for $\mathbb{R}^n$ will serve for the other examples.

Compact Lie groups.

Here we take A to be a two-sided invariant Laplace-Beltrami operator in the group. If we let $X_1, \ldots X_n$, be a basis for the right invariant Lie algebra (considered as right invariant first-order differential operators on G), then we can choose $A = \Sigma\, a_{ij} X_i X_j$ where $\{a_{ij}\}$ is an appropriate positive definite symmetric matrix. Set

$$R_j = X_j(-A)^{-1/2}.$$

Then again $$g_1(R_j f)(x) \leq c\, g(f)(x)$$

and $$\|R_j(f)\|_p \leq A_p \|f\|_p, \qquad 1 < p < \infty.$$

Semi-simple Lie groups

Let G/K be a symmetric pair, with K a maximal compact subgroup of G. We restrict our attention to zonal functions f; these can be considered as functions on G which are right and left invariant under K; the condition is $f(k_1 x k_2) = f(x)$, if $x \in G$, $k_1, k_2 \in K$. We set A to be the restriction of the Casimir operator to smooth zonal functions. Finally we write $R_j = X_j(-A)^{-1/2}$. Then again

$$g_1(R_j f)(x) \leq c\, g(f)(x), \text{ and}$$

$$\|R_j(f)\|_p \leq A_p \|f\|_p, \qquad 1 < p < \infty.$$

As a final remark we should point out that much of this approach also works for the Sturm-Liouville case, (Example 4 in Section 4). A detailed study of the ultraspherical and Fourier-Bessel expansions had been made earlier, more in the spirit of the development of Section 3.

REFERENCES

[0] D. L. Burkholder, Annals of Math. Statistics 37 (1966), 1494-1504.

[1] A. Benedek, A. P. Calderón, and R. Panzone, Proc. Nat. Acad. Sci. U.S.A., 48 (1962), 356-365.

[2] A. P. Calderón, ibid, 53 (1965), 1092-1099.

[3] A. P. Calderón and A. Zygmund, Acta Math 88 (1952), 85-139.

[4] C. Fefferman, Thesis at Princeton University, 1969.

[5] G. Gasper, Proc. Nat. Acad. Sci. U.S.A., 57 (1967), 25-28.

[6] R. Gundy, Ann. Math. Statistics 39 (1968), 134-138.

[7] J. E. Littlewood and R. E. Paley, Proc. London Math. Soc. 42 (1937), 52-89; ibid, 43 (1937), 105-126.

[8] J. Marcinkiewicz, Studia Math 8 (1939), 78-91.

[9] R. E. Paley, Proc. London Math. Soc. 34 (1931), 241-264, 265-279.

[10] C. Segovia, Thesis at University of Chicago, 1967.

[11] E. M. Stein, Trans. Amer. Math. Soc. 88 (1958), 430-466.

[12] _______, C. R. Acad. Sci., Paris, 1966, 1967.

[13] _______, "Intégrales Singulieres, et Fonctions Differentiable de Plusieurs Variables," Lecture Notes at Orsay, France for the academic year 1966-1967, Chapters I-VI published in 1967.

[14] _______, "Topics in Harmonic Analysis related to the Littlewood-Paley Theory," to appear in Annals of Math. Studies.

[15] E. M. Stein and G. Weiss, Acta Math. 103 (1960), 25-62.

[16] N. Wiener, Duke Math. Journal 5 (1939), 1-18.

[17] A. Zygmund, "Trigonometric Series," Cambridge Univ. Press, 1959.

Problems in the Theory of Automorphic Forms

To Salomon Bochner

In Gratitude

R. P. Langlands

1. There has recently been much interest, if not a tremendous amount of progress, in the arithmetic theory of automorphic forms. In this lecture I would like to present the views not of a number theorist but of a student of group representations on those of its problems that he finds most fascinating. To be more precise I want to formulate a series of questions which the reader may, if he likes, take as conjectures. I prefer to regard them as working hypotheses. They have already led to some interesting facts. Although they have stood up for a fair length of time to the most careful scrutiny I could give I am still not entirely easy about them. Indeed even at the beginning in the course of the definitions, which I want to make in complete generality, I am forced, for lack of time and technical competence, to make various assumptions.

I should perhaps apologize for such a speculative lecture. However there are some interesting facts scattered amongst the questions. Moreover the unsolved problems in group representations arising from the theory of automorphic forms are much less technical than the solved ones and their significance can perhaps be more easily appreciated by the outsider.

Suppose G is a connected reductive algebraic group defined over a global field F. F is then an algebraic number field or a function field in one variable over a finite field. Let $\mathbb{A}(F)$ be the adèle ring of F. $G_{\mathbb{A}(F)}$ is a locally compact topological group with G_F as a discrete subgroup. The group $G_{\mathbb{A}(F)}$ acts on the functions on $G_F\backslash G_{\mathbb{A}(F)}$. In particular it acts on $L^2(G_F\backslash G_{\mathbb{A}(F)})$. It should be possible, although I have not done so and it is not important at this stage, to attach a precise meaning to the assertion that a given irreducible representation π of $G_{\mathbb{A}(F)}$ occurs in $L^2(G_F\backslash G_{\mathbb{A}(F)})$.

If G is abelian it would mean that π is a character of $G_F \backslash G_{A(F)}$. If G is not abelian it would be true for at least those representations which act on an irreducible invariant subspace of $L^2(G_F \backslash G_{A(F)})$.

If G is $GL(1)$ then to each such π one, following Hecke, associates an L-function. If G is $GL(2)$ then Hecke has also introduced, without explicitly mentioning group representations, some L-functions. The problems I want to discuss center about the possibility of defining L-functions for all such π and proving that they have the analytic properties we have grown used to expecting of such functions. I shall also comment on the possible relations of these new functions to the Artin L-functions and the L-functions attached to algebraic varieties.

Given G I am going to introduce a complex analytic group $\hat{G}_F$. To each complex analytic representation σ of $\hat{G}_F$ and each π I want to attach an L-function $L(s,\sigma,\pi)$. Let me say a few words about the general way in which I want to form the function. $G_{A(F)}$ is a restricted direct product $\prod_{\mathfrak{p}} G_{F_{\mathfrak{p}}}$. The product is taken over the primes, finite and infinite, of F. It is reasonable to expect although to my knowledge it has not yet been proved in general that π can be represented as $\prod_{\mathfrak{p}} \otimes \pi_{\mathfrak{p}}$ where $\pi_{\mathfrak{p}}$ is a unitary representation of $G_{F_{\mathfrak{p}}}$.

I would like to have first associated to any algebraic group G defined over $F_{\mathfrak{p}}$ a complex analytic group $\hat{G}_{F_{\mathfrak{p}}}$ and to any complex analytic representation $\sigma_{\mathfrak{p}}$ of $\hat{G}_{F_{\mathfrak{p}}}$ and any unitary representation $\pi_{\mathfrak{p}}$ of $G_{F_{\mathfrak{p}}}$ a local L-function $L(s,\sigma_{\mathfrak{p}},\pi_{\mathfrak{p}})$ which, when $\mathfrak{p}$ is non-archimedean, would be of the form

$$\prod_{i=1}^{n} \frac{1}{1 - \alpha_i |\Pi_{\mathfrak{p}}|^s}$$

where n is the degree of $\sigma_\wp$. Some of the α_i may be zero. For $\wp$ infinite it would be, basically, a product of Γ-functions. $L(s,\sigma_\wp,\pi_\wp)$ would depend only on the equivalence classes of $\sigma_\wp$ and $\pi_\wp$. I would also like to have defined for every non-trivial additive character $\Psi_{F_\wp}$ of $F_\wp$ a factor $\varepsilon(s,\sigma_\wp,\pi_\wp,\Psi_{F_\wp})$ which, as a function of s , has the form ae^{bs} .

There would be a complex analytic homomorphism of $\hat{G}_{F_\wp}$ into $\hat{G}_F$ determined up to an inner automorphism of $\hat{G}_F$. Thus σ determines for each $\wp$ a representation $\sigma_\wp$ of $\hat{G}_{F_\wp}$. I want to define

$$L(s,\sigma,\pi) = \prod_\wp L(s,\sigma_\wp,\pi_\wp) \ . \tag{A}$$

Of course it has to be shown that the product converges in a half-plane. We shall see how to do this. Then we will want to prove that the function can be analytically continued to a function meromorphic in the whole complex plane. Let Ψ_F be a non-trivial character of $F \backslash A(F)$ and let $\Psi_{F_\wp}$ be the restriction of Ψ_F to $F_\wp$. We will want $\varepsilon(s,\sigma_\wp,\pi_\wp,\Psi_{F_\wp})$ to be 1 for all but finitely many $\wp$. We will also want

$$\varepsilon(s,\sigma,\pi) = \prod_\wp \varepsilon(s,\sigma_\wp,\pi_\wp,\Psi_{F_\wp})$$

to be independent of Ψ_F . The functional equation should be

$$L(s,\sigma,\pi) = \varepsilon(s,\sigma,\pi) \ L(1-s,\tilde{\sigma},\pi)$$

if $\tilde{\sigma}$ is the representation contragredient to σ .

We are asking for too much too soon. What we should try to do is to define the $L(s,\sigma_\wp,\pi_\wp)$ and the $\varepsilon(s,\sigma_\wp,\pi_\wp,\Psi_{F_\wp})$ when there is no ramification, verify that there is ramification at only a finite number of primes, and show that if the product in (A) is taken only over the unramified primes it converges for Re s sufficiently large. As we learn how to prove the functional equations we shall be able to make the definitions at the unramified primes. By the way we introduce the additive characters, whose appearance must appear rather mysterious, only because we can indeed prove some things and know better than to leave them out.

What does unramified mean in our context? First of all for $\mathfrak{f}$ to be unramified G will have to be quasi-split over $F_{\mathfrak{f}}$ and split over an unramified extension. In that case there is, as we shall see, a canonical conjugacy class of maximal compact subgroups of $G_{F_{\mathfrak{f}}}$. For $\mathfrak{f}$ to be unramified the restriction of $\pi_{\mathfrak{f}}$ to any one of these groups will have to contain the identity representation. There is also a condition to be imposed on $\Psi_{F_{\mathfrak{f}}}$. Although it is not very important I would like to mention it explicitly. If $\mathfrak{f}$ is non-archimedean the largest ideal of $F_{\mathfrak{f}}$ on which $\Psi_{F_{\mathfrak{f}}}$ is trivial will have to be $O_{F_{\mathfrak{f}}}$, the ring of integers in $F_{\mathfrak{f}}$. If $F_{\mathfrak{f}}$ is $\mathbb{R}$ then $\Psi_{F_{\mathfrak{f}}}(x)$ will have to be $e^{2\pi ix}$ and if $F_{\mathfrak{f}}$ is $\mathbb{C}$ then $\Psi_{F_{\mathfrak{f}}}(z)$ will have to be $e^{4\pi i \operatorname{Re} z}$. We want $\varepsilon(s,\sigma_{\mathfrak{f}},\pi_{\mathfrak{f}},\Psi_{F_{\mathfrak{f}}})$ to be 1 if $\mathfrak{f}$ is unramified.

2. $\hat{G}_F$ can be defined for a connected reductive group over any field F. Take first a quasi-split group G over F which splits over the Galois extension K. Choose a Borel subgroup B of G which is defined over F and let T be a maximal torus of B which is also defined over F. Let L be the group of rational characters of T. Write G as G^oG^1 where G^o is abelian and G^1 is semi-simple. Then $G^o \cap G^1$ is finite. If $T^o = G^o$ and $T^1 = T \cap G^1$ then $T = T^oT^1$. Let L_+^o be the group of rational characters of T^o and let L_-^o be the elements of L_+^o which are 1 on $T^o \cap T^1$. Let L_-^+ be the group generated by the roots of T^1. If R is any field let

$E^1_R = L^1_- \otimes_{\mathbb{Z}} R$. The Weyl group Ω acts on L^1_- and therefore on E^1_R . Let $(\cdot, \cdot)$ be a non-degenerate bilinear form on $E^1_{\mathbb{C}}$ which is invariant under Ω . Suppose also that its restriction to $E^1_{\mathbb{R}}$ is positive definite. Let

$$L^1_+ = \{\lambda \in E^1_{\mathbb{C}} \mid 2\frac{(\lambda,\alpha)}{(\alpha,\alpha)} \in \mathbb{Z} \text{ for all roots } \alpha\} .$$

Set $L_- = L^o_- \oplus L^1_-$ and $L_+ = L^o_+ \oplus L^1_+$. We may regard L as a sublattice of L_+ . It will contain L_- .

Let $\alpha_1, \ldots, \alpha_\ell$ be the simple roots of T^1 with respect to B and let

$$(A_{ij}) = 2\frac{(\alpha_i,\alpha_j)}{(\alpha_i,\alpha_i)}$$

be the Cartan matrix. If σ belongs to $\mathfrak{G}(K/F)$ and λ belongs to L then $\sigma\lambda$, where $\sigma\lambda(t) = \sigma(\lambda(\sigma^{-1}t))$, also belongs to L . Thus $\mathfrak{G}(K/F)$ acts on L . It also acts on L_- and L_+ and the actions on these three lattices are consistent. Moreover the roots $\alpha_1, \ldots, \alpha_\ell$ are permuted amongst themselves and the Cartan matrix is left invariant.

If R is any field containing $\mathbb{Q}$ let $E_R = L \otimes_{\mathbb{Z}} R$ and let $\hat{E}_R = \mathrm{Hom}_R(E_R, R)$. The lattices

$$\hat{L}_+ = \mathrm{Hom}(L_-, \mathbb{Z}) = \mathrm{Hom}(L^o_-, \mathbb{Z}) \oplus \mathrm{Hom}(L^1_-, \mathbb{Z}) = \hat{L}^o_+ \oplus \hat{L}^1_+$$

$$\hat{L} = \mathrm{Hom}(L, \mathbb{Z})$$

$$\hat{L}_- = \mathrm{Hom}(L_+, \mathbb{Z}) = \mathrm{Hom}(L^o_+, \mathbb{Z}) \oplus \mathrm{Hom}(L^1_+, \mathbb{Z}) = \hat{L}^o_- \oplus \hat{L}^1_-$$

may be regarded as subgroups of $\hat{E}_{\mathbb{C}}$. If $E^o_R = L^o_- \otimes_{\mathbb{Z}} R$ then $E_R = E^o_R \oplus E^1_R$. With the obvious definitions of $\hat{E}^o_R$ and $\hat{E}^1_R$ we have $\hat{E}_R = \hat{E}^o_R \oplus \hat{E}^1_R$. Let

$(\cdot\,,\cdot)$ also denote the form on $\hat{E}^1_{\mathbb{C}}$ adjoint to the given form on $E^1_{\mathbb{C}}$. To be precise if λ and μ belong to $E^1_{\mathbb{C}}$, if $\hat{\lambda}$ and $\hat{\mu}$ belong to $\hat{E}^1_{\mathbb{C}}$, and if $\langle\eta,\hat{\lambda}\rangle = (\eta,\lambda)$ and $\langle\eta,\hat{\mu}\rangle = (\eta,\mu)$ for all η in $E^1_{\mathbb{C}}$ then $(\lambda,\mu) = (\hat{\lambda},\hat{\mu})$.

If α is a root define its coroot $\hat{\alpha}$ in $\hat{E}^1_{\mathbb{C}}$ by the condition:

$$\langle\lambda,\hat{\alpha}\rangle = 2\,\frac{(\lambda,\alpha)}{(\alpha,\alpha)}$$

for all λ in $E^1_{\mathbb{C}}$. The coroots generate $\hat{L}^1_-$. Moreover

$$(\hat{\alpha},\hat{\beta}) = 4\,\frac{(\alpha,\beta)}{(\alpha,\alpha)(\beta,\beta)}$$

and

$$2\,\frac{(\hat{\alpha},\hat{\beta})}{(\hat{\alpha},\hat{\alpha})} = 2\,\frac{(\alpha,\beta)}{(\beta,\beta)}\,.$$

Thus the matrix

$$(\hat{A}_{ij}) = \left(2\,\frac{(\hat{\alpha}_i,\hat{\alpha}_j)}{(\hat{\alpha}_i,\hat{\alpha}_i)}\right)$$

is the transpose of (A_{ij}). The linear transformation $\hat{S}_i$ of $\hat{E}^1_{\mathbb{C}}$ defined by

$$\hat{S}_i(\hat{\alpha}_j) = \hat{\alpha}_j - \hat{A}_{ij}\,\hat{\alpha}_i = \hat{\alpha}_j - A_{ji}\hat{\alpha}_i$$

is contragredient to the linear transformation S_i of $E^1_{\mathbb{C}}$ defined by

$$S_i(\alpha_j) = \alpha_j - A_{ij}\alpha_i\,.$$

Thus the group $\hat{\Omega}$ generated by $\{\hat{S}_i \mid 1 \leq i \leq \ell\}$ is canonically isomorphic to the finite group Ω and, by a well-known theorem (cf. Chapter VII of [7]) $(\hat{A}_{ij})$ is the Cartan matrix of a simply-connected complex group $\hat{G}^1_+$. Let $\hat{B}^1_+$ be a Borel subgroup of $\hat{G}^1_+$ and let $\hat{T}^1_+$ be a Cartan subgroup in $\hat{B}^1_+$. We identify the simple roots of $\hat{T}^1_+$ with respect to $\hat{B}^1_+$ with $\hat{\alpha}_1, \ldots, \hat{\alpha}_\ell$ and the free vector space over $\mathbb{C}$ with basis $\{\hat{\alpha}_1, \ldots, \hat{\alpha}_\ell\}$ with $\hat{E}^1_{\mathbb{C}}$. We may also identify Ω and $\hat{\Omega}$. The roots of $\hat{T}^1_+$ are the vectors $\omega\hat{\alpha}_i$, $\omega \in \Omega$, $1 \leq i \leq \ell$. If $\omega\alpha_i = \alpha$ then $\omega\hat{\alpha}_i = \hat{\alpha}$ because

$$\langle \lambda, \omega\hat{\alpha}_i \rangle = \langle \omega^{-1}\lambda, \hat{\alpha}_i \rangle = 2\frac{(\omega^{-1}\lambda, \alpha_i)}{(\alpha_i, \alpha_i)} = 2\frac{(\lambda, \omega\alpha_i)}{(\omega\alpha_i, \omega\alpha_i)} = 2\frac{(\lambda, \alpha)}{(\alpha, \alpha)}$$

Thus the roots of $\hat{T}^1_+$ are just the coroots. If λ belongs to $\hat{E}^1_{\mathbb{C}}$ then

$$2\frac{(\lambda, \hat{\alpha})}{(\hat{\alpha}, \hat{\alpha})} = \langle \alpha, \lambda \rangle$$

so that

$$\hat{L}^1_+ = \{\lambda \in \hat{E}^1_{\mathbb{C}} \mid 2\frac{(\lambda, \hat{\alpha})}{(\hat{\alpha}, \hat{\alpha})} \in \mathbb{Z} \text{ for all coroots } \hat{\alpha}\}$$

and is therefore just the set of weights of $\hat{T}^1_+$.

Let

$$\hat{G}^o_+ = \mathrm{Hom}_{\mathbb{Z}}(\hat{L}^o_+, \mathbb{C}^*) .$$

$\hat{G}^o_+$ is a reductive complex Lie group. Set $\hat{G}_+ = \hat{G}^o_+ \times \hat{G}^1_+$. If $\hat{T}^o_+ = \hat{G}^o_+$ and $\hat{T}_+ = \hat{T}^o_+ \times \hat{T}^1_+$ then $\hat{L}_+$ is the set of complex analytic characters of $\hat{T}_+$. If

$$\hat{Z} = \{t \in \hat{T}_+ \mid \lambda(t) = 1 \quad \text{for all } \lambda \text{ in } \hat{L}\}$$

then $\hat{Z}$ is a normal subgroup of $\hat{G}_+$ and $\hat{G} = \hat{G}_+ / \hat{Z}$ is also a complex Lie group. $\mathfrak{G}(K/F)$ acts in a natural fashion on $\hat{L}_-$, $\hat{L}$, and $\hat{L}_+$. The action leaves the set $\{\hat{\alpha}_1, \ldots, \hat{\alpha}_\ell\}$ invariant. $\mathfrak{G}(K/F)$ acts naturally on $\hat{G}^o_+$. I want to define an action on $\hat{G}^1_+$ and therefore an action on $\hat{G}_+$. Choose $H_1, \ldots, H_\ell$ in the Lie algebra of $\hat{T}^1_+$ so that

$$\lambda(H_i) = \langle \alpha_i, \lambda \rangle$$

for all λ in $\hat{L}^1_+$. Choose root vectors $X_1, \dots, X_\ell$ belonging to the coroots $\hat{\alpha}_1, \dots, \hat{\alpha}_\ell$ and root vectors $Y_1, \dots, Y_\ell$ belonging to their negatives. Suppose $[X_i, Y_i] = H_i$. If σ belongs to $\mathfrak{G}(K/F)$ let $\sigma(\hat{\alpha}_i) = \hat{\alpha}_{\sigma(i)}$. There is (cf. Chapter VII of [7]) a unique isomorphism σ of the Lie algebra of $\hat{G}^1_+$ so that

$$\sigma(H_i) = H_{\sigma(i)} \, , \; \sigma(X_i) = X_{\sigma(i)} \, , \; \sigma(Y_i) = Y_{\sigma(i)} \, .$$

These isomorphisms clearly determine an action of $\mathfrak{G}(K/F)$ on the Lie algebra and therefore one on $\hat{G}^1_+$ itself. Since $\mathfrak{G}(K/F)$ leaves L invariant its action on $\hat{G}_+$ can be transferred to $\hat{G}$. If $\hat{B}$ is the image of $\hat{B}_+ = \hat{T}^0_+ \times \hat{B}^1_+$ and $\hat{T}$ the image of $\hat{T}_+$ in $\hat{G}$ the action leaves $\hat{B}$ and $\hat{T}$ invariant. I want to define $\hat{G}_F$ to be the semi-direct product $\hat{G} \times \mathfrak{G}(K/F)$.

However $\hat{G}_F$ as defined depends upon the choice of B, T, and $X_1, \dots, X_\ell$ and $\hat{G}_F$ comes provided with a Borel subgroup $\hat{B}$ of its connected component, a Cartan subgroup $\hat{T}$ of $\hat{B}$, and a one-to-one correspondence between the simple roots of T with respect to B and those of $\hat{T}$ with respect to $\hat{B}$. Suppose G' is another quasi-split group over F which is isomorphic to G over K by means of an isomorphism φ such that $\varphi^{-1}\sigma(\varphi)$ is inner for all σ in $\mathfrak{G}(K/F)$, B' is a Borel subgroup of G' defined over F , and T' is a Cartan subgroup of B' also defined over F . There is an inner automorphism Ψ of G which is defined over K so that $\varphi\Psi$ takes B to B' and T to T' . $\varphi\Psi$ determines an isomorphism of $\hat{L}$ and $\hat{L}'$ and a one-to-one correspondence between $\{\alpha_1, \dots, \alpha_\ell\}$ and $\{\alpha'_1, \dots, \alpha'_\ell\}$ both of which depend only on φ and, as is easily verified, commute with the action of $\mathfrak{G}(K/F)$. There is then a natural isomorphism of $\hat{G}^0_+$ with $(\hat{G}^0_+)'$ associated to φ . Moreover there is a unique isomorphism of $\hat{G}^1_+$ with $(\hat{G}^1_+)'$

whose action on the Lie algebras takes H_i to H'_i , X_i to X'_i , and Y_i to Y'_i . The two together define an isomorphism of $\hat{G}_+$ with $\hat{G}'_+$. If we assume that α_i corresponds to α'_i , $1 \leq i \leq \ell$ this isomorphism takes $\hat{Z}$ to $\hat{Z}'$ and determines an isomorphism of $\hat{G}$ with $\hat{G}'$ which commutes with $\mathfrak{G}(K/F)$. This in turn determines an isomorphism $\hat{\varphi}$ of $\hat{G}'_F$ with $\hat{G}_F$. In particular taking $G' = G$ and φ to be the identity we see that $\hat{G}_F$ is determined up to a canonical isomorphism.

Suppose G is any reductive group over F , K is a Galois extension of F , G' and G'' are quasi-split groups over F which split over K , and $\varphi : G' \longrightarrow G$, $\Psi : G'' \longrightarrow G$ are isomorphisms defined over K such that $\varphi^{-1}\sigma(\varphi)$ and $\Psi^{-1}\sigma(\Psi)$ are inner for all σ in $\mathfrak{G}(K/F)$. Then $(\Psi^{-1}\varphi)^{-1}\sigma(\Psi^{-1}\varphi)$ is also inner so that there is a canonical isomorphism of $\hat{G}'_F$ and $\hat{G}''_F$. We are thus free to set $\hat{G}_F = \hat{G}'_F$. $\hat{G}_F$ depends on K but there is no need to stress this. However we shall sometimes write $\hat{G}_{K/F}$ instead of $\hat{G}_F$.

3. Although it is a rather simple case it may be worthwhile to carry out the previous construction when G is $GL(n)$ and $K = F$. We take T to be the diagonal and B to be the upper triangular matrices. G^0 is the group of non-zero scalar matrices and G^1 is $SL(n)$. If λ belongs to L and

$$\lambda : \begin{pmatrix} t_1 & & o \\ & \ddots & \\ o & & t_n \end{pmatrix} \longrightarrow t_1^{m_1} \cdots t_n^{m_n}$$

with $m_1 , \ldots , m_n$ in $\mathbb{Z}$ we write $\lambda = (m_1, \ldots , m_n)$. Thus L is identified with $\mathbb{Z}^n$. We may identify $E_{\mathbb{R}}$ with $\mathbb{R}^n$ and $E_{\mathbb{C}}$ with $\mathbb{C}^n$. If λ belongs to L^0_+ and

$$\lambda : tI \longrightarrow t^m$$

with m in $\mathbb{Z}$ we write $\lambda = \left(\frac{m}{n}, \ldots, \frac{m}{n}\right)$. Then L_-^o which is a subgroup of both L and L_+^o consists of the elements $(m, \ldots, m)$ with m in $\mathbb{Z}$. The rank ℓ is $n-1$ and

$$\alpha_1 = (1, -1, 0, \ldots, 0)$$

$$\alpha_2 = (0, 1, -1, 0, \ldots, 0)$$

$$\vdots$$

$$\alpha_\ell = (0, \ldots, 0, 1, -1)$$

Thus

$$L_-^1 = \{(m_1, \ldots, m_n) \in L \mid \textstyle\sum_{i=1}^n m_i = 0\} .$$

$E_{\mathbb{C}}^1$ is the set of all $(z_1, \ldots, z_n)$ in $E_{\mathbb{C}}$ for which

$$\textstyle\sum_{i=1}^n z_i = 0 .$$

The bilinear form on $E_{\mathbb{C}}^1$ may be taken as the restriction of the form

$$(z,w) = \textstyle\sum_{i=1}^n z_i w_i$$

on $E_{\mathbb{C}}$. Thus

$$L_+^1 = \{(m_1, \ldots, m_n) \mid \textstyle\sum_{i=1}^n m_i = 0 \text{ and } m_i - m_j \in \mathbb{Z}\} .$$

We may use the given bilinear form to identify $\hat{E}_{\mathbb{C}}$ with $E_{\mathbb{C}}$. Then the $\hat{}$-operation leaves all lattices and all roots fixed. Thus $\hat{G}_+^o = \mathrm{Hom}(L_+^o, \mathbb{C})$. Any non-singular complex scalar matrix tI defines an element of $\hat{G}_+^o$, namely, the homomorphism

$$\left(\frac{m}{n}, \ldots, \frac{m}{n}\right) \longrightarrow t^m .$$

We identify $\hat{G}^o_+$ with the group of scalar matrices. $\hat{G}^1_+$ is $SL(n,\mathbb{C})$. There is a natural map of $\hat{G}^o_+ \times \hat{G}^1_+$ onto $GL(n,\mathbb{C})$. It sends $tI \times A$ to tA. The kernel is easily seen to be $\hat{Z}$ so that $\hat{G}_F$ is $GL(n,\mathbb{C})$.

4. To define the local L-functions, to prove that almost all primes are unramified, and to prove that the product of the local L-functions over the unramified primes converges for $\operatorname{Re} s$ sufficiently large we need some facts from the reduction theory for groups over local fields (cf. [1]). Much progress has been made in that theory but it is still incomplete. Unfortunately the particular facts we need do not seem to be in the literature. Very little is lost at this stage if we just assume them. For the groups about which something definite can be said they are easily verified.

Suppose K is an unramified extension of the non-archimedean local field F and G is a quasi-split group over F which splits over K. Let B be a Borel subgroup of G and T a Cartan subgroup of B both of which are defined over F. Let v be the valuation on K. It is a homomorphism from K^*, the multiplicative group of K, onto $\mathbb{Z}$ whose kernel is the group of units. If t belongs to T_F let $v(t)$ in $\hat{L}$ be defined by $\langle\lambda,v(t)\rangle = v(\lambda(t))$ for all λ in L. If σ belongs to $\mathfrak{G}(K/F)$ then

$$\langle\lambda,\sigma v(t)\rangle = \langle\sigma^{-1}\lambda,v(t)\rangle = v(\sigma^{-1}(\lambda(\sigma t))) = v(\lambda(t))$$

because $\sigma t = t$ and $v(\sigma^{-1}a) = v(a)$ for all a in K^*. Thus v is a homomorphism of T_F into $\hat{M}$, the group of invariants of $\mathfrak{G}(K/F)$ in $\hat{L}$. It is in fact easily seen that it takes T_F onto $\hat{M}$.

We assume the following lemma.

Lemma 1. *There is a Chevalley lattice in the Lie algebra of G whose stabilizer U_K is invariant under $\mathfrak{G}(K/F)$. U_K is its own normalizer. Moreover $G_K = B_K U_K$, $H^1(\mathfrak{G}(K/F), U_K) = 1$, and $H^1(\mathfrak{G}(K/F), B_K \cap U_K) = 1$. If we choose two such Chevalley lattices with stabilizers U_K and U_K' respectively then U_K' is conjugate to U_K in G_K.*

If g belongs to G_K and σ belongs to $\mathfrak{G}(K/F)$ let $g^\sigma = \sigma^{-1}(g)$. If g belongs to G_F we may write it as $g = bu$ with b in B_K and u in U_K. Then $g^\sigma = b^\sigma u^\sigma$ and $u^\sigma u^{-1} = b^{-\sigma} b$. By the lemma there is a v in $B_K \cap U_K$ so that $u^\sigma u^{-1} = b^{-\sigma} b = v^\sigma v^{-1}$. Then $b' = bv$ belongs to B_F, $u' = v^{-1}u$ belongs to $U_F = G_F \cap U_K$, and $g = b'u'$. Thus $G_F = B_F U_F$.

If $gU_K g^{-1} = U_K'$ for some g in G_K then $g^\sigma U_K g^{-\sigma} = U_K'$ so that $g^{-\sigma} g$ belongs to U_K which is its own normalizer. By the lemma there is u in U_K so that $g^{-\sigma} g = u^\sigma u^{-1}$. Then $g_1 = gu$ lies in G_F and $g_1 U_K g_1^{-1} = U_K'$. Thus U_F and U_F' are conjugate in G_F.

Let $C_c(G_F, U_F)$ be the set of all compactly supported functions for G_F such that $f(gu) = f(ug) = f(g)$ for all u in U_F and all g in G_F. $C_c(G_F, U_F)$ is an algebra under convolution. It is called the Hecke algebra. If N is the unipotent radical of B let dn be a Haar measure on N_F and let $\frac{d(bnb^{-1})}{dn} = \delta(b)$ if b belongs to B_F. If λ belongs to $\hat{M}$ choose t in T_F so that $v(t) = \lambda$. If f belongs to $C_c(G_F, U_F)$ set

$$\hat{f}(\lambda) = \delta^{1/2}(t) \left\{ \int_{N_F \cap U_F} dn \right\}^{-1} \int_{N_F} f(tn)\,dn .$$

The group $\mathfrak{G}(K/F)$ acts on Ω . Let Ω^o be the group of invariant elements. Ω^o acts on $\hat{M}$. Let $\Lambda(\hat{M})$ be the group algebra of $\hat{M}$ over $\mathbb{C}$ and let $\Lambda^o(\hat{M})$ be the invariants of Ω^o in $\Lambda(\hat{M})$. We also assume the following lemma (cf. [12]).

<u>Lemma 2</u>. <u>The map</u> $f \longrightarrow \hat{f}$ <u>is an isomorphism of</u> $C_c(G_F,U_F)$ <u>and</u> $\Lambda^o(M)$.

Suppose B is replaced by B_1 and T by T_1 . Observe that $T \simeq B/N$ and $T_1 \simeq B_1/N_1$. If u in G_F takes B to B_1 it takes N to N_1 and defines a map from T to T_1 . This map does not depend on u . It determines $\mathfrak{G}(K/F)$ invariant maps from L_1 to L and from $\hat{L}$ to $\hat{L}_1$ and thus maps from $\hat{M}$ to $\hat{M}_1$ and from $\Lambda^o(\hat{M})$ to $\Lambda^o(\hat{M}_1)$. Suppose $\hat{f}$ goes to $\hat{f}_1$ and $\hat{\lambda}$ goes to $\hat{\lambda}_1$. If we choose, as we may, u in U_F then

$$\hat{f}_1(\hat{\lambda}_1) = \hat{f}(\hat{\lambda}) = \delta^{1/2}(t)\ \{\int_{N_F \cap U_F} dn\}^{-1} \int_{N_F} f(tn)dn \quad .$$

Let $N_F \cap U_F = V$. Denote the corresponding group associated to N_1 by V_1 . Then $u\,Vu^{-1} = V_1$. Choose $d(unu^{-1}) = dn_1$. Since $f(ugu^{-1}) = f(g)$ the expression on the right equals

$$\delta^{1/2}(utu^{-1})\ \{\int_{V_1} dn_1\}^{-1} \int_{N_F} f(utu^{-1}unu^{-1})dn \quad .$$

If utu^{-1} projects on t_1 in T_1 then $\delta(utu^{-1}) = \delta(t_1)$ and $v(t_1) = \hat{\lambda}_1$. Moreover

$$\int f(utu^{-1}unu^{-1})dn = \int f(t_1n_1)dn_1$$

and the diagram

$$\begin{array}{ccc} & C_c(G_F,\ U_F) & \\ \swarrow & & \searrow \\ \Lambda^o(M) & \longrightarrow & \Lambda^o(M_1) \end{array}$$

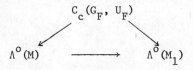

is commutative.

If $gU_Fg^{-1} = U_F'$ the map $f \longrightarrow f'$ with $f'(h) = f(g^{-1}hg)$ is an isomorphism of $C_c(G_F,U_F)$ with $C_c(G_F,U_F')$. It does not depend on g . We can take g in B_F . Then

$$\hat{f}'(\lambda) = \delta^{1/2}(t)\ \{\int_{N_F \cap U_F'} dn\}^{-1} \int_{N_F} f(g^{-1}tng)dn\ .$$

Since $g^{-1}tng = t(t^{-1}g^{-1}tg)g^{-1}ng$ the second integral is equal to

$$\int_{N_F} f(tg^{-1}ng)dn\ .$$

Since

$$\frac{d(g^{-1}ng)}{dn} = \{\int_{N_F \cap U_F'} dn\}^{-1} \int_{N_F \cap U_F} dn$$

we conclude that $\hat{f}'(\hat{\lambda}) = \hat{f}(\hat{\lambda})$ and that the diagram

$$\begin{array}{ccc} C_c(G_F,U_F) & \longrightarrow & C_c(G_F,U_F') \\ & \searrow \quad \swarrow & \\ & \Lambda^o(\hat{M}) & \end{array}$$

is commutative.

I shall not explicitly mention the commutativity of these diagrams again. However they are important because they imply that the definitions to follow have the invariance properties which are required if they are to have any sense.

If π is an irreducible unitary representation of G_F on H whose restriction to U_F contains the identity representation then

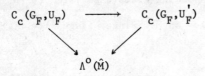

$$H_o = \{ x \in H \mid \pi(u)\, x = x \quad \text{for all} \quad u \quad \text{in} \quad U_F \}$$

is a one-dimensional subspace. If f belongs to $C_c(G_F, U_F)$ then

$$\pi(f) = \int_G f(g)\pi(g)dg$$

maps H_o into itself. The representation of $C_c(G_F,U_F)$ on H_o determines a homomorphism χ of $C_c(G_F,U_F)$ or of $\Lambda^o(\hat{M})$ into the ring of complex numbers. π is determined by χ. To define the local L-functions we study such homomorphisms. First of all observe that if χ is associated to a unitary representation then

$$|\chi(f)| \leq \int_{G_F} |f(g)|dg \; .$$

Since $\Lambda(\hat{M})$ is a finitely generated module over $\Lambda^o(\hat{M})$ any homomorphism of $\Lambda^o(\hat{M})$ into $\mathbb{C}$ may be extended to a homomorphism of $\Lambda(\hat{M})$ into $\mathbb{C}$ which will necessarily be of the form

$$\Sigma \hat{f}(\lambda)\lambda \longrightarrow \Sigma \hat{f}(\lambda)\lambda(t) \tag{B}$$

for some t in $\hat{T}$. Conversely given t the formula (B) determines a homomorphism χ_t of $\Lambda^o(\hat{M})$ into $\mathbb{C}$. We shall show that $\chi_{t_1} = \chi_{t_2}$ if and only if $t_1 \times \sigma_F$ and $t_2 \times \sigma_F$, where σ_F is the Frobenius substitution, are conjugate in $\hat{G}_F$. If t belongs to $\hat{G}$ and σ belongs to $\mathfrak{G}(K/F)$ we shall abbreviate $t \times \sigma$ to $t\,\sigma$. It is known [4] that every semi-simple element of $\hat{G}_F$ whose projection on $\mathfrak{G}(K/F)$ is σ_F is conjugate to some $t\,\sigma_F$ with t in $\hat{T}$. Thus there is a one-to-one correspondence between homomorphisms of the Hecke algebra into $\mathbb{C}$ and semi-simple conjugacy classes in $\hat{G}_F$ whose projection on $\mathfrak{G}(K/F)$ is σ_F.

If ρ is a complex analytic representation of $\hat{G}_F$ and χ_t is the homomorphism of $\Lambda^o(\hat{M})$ into $\mathbb{C}$ associated to π we define the local L-function to be

$$L(s,\rho,\pi) = \frac{1}{\det(I-\rho(t\sigma_F)|\pi_F|^s)}$$

if π_F generates the maximal ideal of O_F.

$\hat{T}$ may be identified with $\mathrm{Hom}_{\mathbb{Z}}(\hat{L},\mathbb{C}^*)$. The exact sequence

$$0 \longrightarrow \mathbb{Z} \xrightarrow{\varphi} \mathbb{C} \xrightarrow{\Psi} \mathbb{C}^* \longrightarrow 0$$

with $\varphi(z) = \frac{2\pi i}{\log|\pi_F|} z$ and $\Psi(z) = |\pi_F|^{-z}$ leads to the exact sequence

$$0 \longrightarrow L = \mathrm{Hom}_{\mathbb{Z}}(\hat{L}, \mathbb{Z}) \xrightarrow{\varphi} E_{\mathbb{C}} = \mathrm{Hom}_{\mathbb{Z}}(\hat{L}, \mathbb{C}) \xrightarrow{\Psi} \hat{T} \longrightarrow 0 .$$

Let $V_{\mathbb{C}}$ be the invariants of $\mathfrak{G}(K/F)$ in $E_{\mathbb{C}}$ and let $W_{\mathbb{C}}$ be the range of $\sigma_F - 1$. Then $E_{\mathbb{C}} = V_{\mathbb{C}} \oplus W_{\mathbb{C}}$. If w belongs to $W_{\mathbb{C}}$ and λ belongs to $\hat{M}$ then $\langle w,\lambda\rangle = 0$ and replacing t by $t\Psi(w)$ does not change χ_t. If $w = \sigma_F v - v$ and $\Psi(v) = s$ then

$$t\Psi(w)\sigma_F = ts^{-1}\sigma_F(s)\sigma_F = s^{-1}(t\sigma_F)s$$

is conjugate to $t\sigma_F$. Thus we have to show that if $t_1 = \Psi(v_1)$ and $t_2 = \Psi(v_2)$ with v_1 and v_2 in $V_{\mathbb{C}}$ then $t_1\sigma_F$ and $t_2\sigma_F$ are conjugate if and only if $\chi_{t_1} = \chi_{t_2}$.

Some preliminary remarks are necessary. We also have a decomposition $\hat{E}_{\mathbb{C}} = \hat{V}_{\mathbb{C}} \oplus \hat{W}_{\mathbb{C}}$ and $\hat{M} = \hat{L} \cap \hat{V}_{\mathbb{C}}$. Let $\hat{Q}$ be the elements of $\hat{V}_{\mathbb{C}}$ obtained by projecting the positive coroots on $\hat{V}_{\mathbb{C}}$. If S is an orbit of $\mathfrak{G}(K/F)$ in the set of positive coroots every element in S has the same projection on $\hat{V}_{\mathbb{C}}$. Since $\Sigma_{\hat{\alpha}\in S}\hat{\alpha}$ belongs to $\hat{V}_{\mathbb{C}}$ the projection must be

$$\frac{1}{n(S)} \Sigma_{\hat{\alpha} \in S} \hat{\alpha}$$

if $n(S)$ is the number of elements in S. Let $S_1, \dots, S_m$ be the orbits of $\mathfrak{G}(K/F)$ in $\{\hat{\alpha}_1, \dots, \hat{\alpha}_\ell\}$ and set

$$\hat{\beta}_i = \frac{1}{n(S_i)} \Sigma_{\hat{\alpha} \in S_i} \hat{\alpha} .$$

Every element of $\hat{Q}$ is a linear combination of $\hat{\beta}_1, \dots, \hat{\beta}_m$ with non-negative coefficients. Notice that if ω belongs to Ω^o and ω acts trivially on $\hat{M}$ then ω leaves each β_i fixed and therefore takes positive roots to positive roots. Thus it is 1. If we extend the inner product in any way from $\hat{E}^1_{\mathbb{R}}$ to $\hat{E}_{\mathbb{R}}$ and set

$$\hat{C} = \{x \in \hat{V}_{\mathbb{R}} | (\hat{\beta}_i, x) \geq 0 , \quad 1 \leq i \leq m\}$$

and

$$\hat{D} = \{x \in \hat{E}_{\mathbb{R}} | (\hat{\alpha}_i, x) \geq 0 , \quad 1 \leq i \leq \ell\}$$

then $\hat{C} = \hat{D} \cap \hat{V}_{\mathbb{R}}$. Consequently no two elements of $\hat{C}$ belong to the same orbit of Ω^o.

Let $\hat{\mathfrak{g}}_i$ be the subalgebra of the Lie algebra of $\hat{G}$ generated by the root vectors belonging to the coroots in S_i and their negatives. $\hat{\mathfrak{g}}_i$ is fixed by $\mathfrak{G}(K/F)$. Let $\hat{G}_i$ be the corresponding analytic group and let $\hat{T}_i = \hat{T} \cap \hat{G}_i$. Let μ_i be the unique element of the Weyl group of $\hat{T}_i$ which takes every positive root to a negative root. If σ belongs to $\mathfrak{G}(K/F)$ then $\sigma(\mu_i)$ has the same property so that $\sigma(\mu_i) = \mu_i$. Let w be any element in the normalizer of $\hat{T}$ whose image in $\hat{\Omega}$ is μ_i. Then $w\sigma_F(w^{-1})$ lies in $\hat{T}$. Its image in $\hat{T}/\Psi(W_{\mathbb{C}})$ is independent of w. I claim that this image is 1.

To see this write $\hat{\mathfrak{g}}_i$ as a direct sum $\sum_{k=1}^{n_i} \hat{\mathfrak{g}}_{ik}$ of simple algebras. If $[K:F] = n$ the stabilizer of $\hat{\mathfrak{g}}_{i1}$ is $\{\sigma_F^{jn_i} | 0 \leq j \leq \frac{n}{n_i}\}$. We may suppose that

$$\hat{\mathfrak{g}}_{ik} = \sigma_F^{k-1} (\hat{\mathfrak{g}}_{i1}) .$$

If $\hat{G}_{ik}$ is the analytic subroup of $\hat{G}$ with Lie algebra $\hat{\mathfrak{g}}_{ik}$ choose w_1 in the normalizer of $\hat{T} \cap \hat{G}_{i1}$ so that w_1 takes the positive roots of $\hat{\mathfrak{g}}_{i1}$ to the negative roots. We may choose w to be $\prod_{k=0}^{n_i-1} \sigma_F^k(w_1)$. Then

$$w\sigma_F(w^{-1}) = (w_1\sigma_F(w_1^{-1}))(\sigma_F(w_1)\sigma_F^2(w_1^{-1}))\ldots(\sigma_F^{n_i-1}(w_1)\sigma_F^{n_i}(w_1^{-1}))$$
$$= w_1 \sigma_F^{n_i}(w_1^{-1}) .$$

The Dynkin diagram of $\hat{\mathfrak{g}}_{i1}$ is connected and the stabilizer of $\hat{\mathfrak{g}}_{i1}$ in $\mathfrak{G}(K/F)$ acts transitively on it. This means that it is of type A_1 or A_2.

In the first case the diagram reduces to a point and the action of the stabilizer must be trivial so that $w_1 = \sigma_F^{n_i}(w_1)$. In the second case $SL(3,\mathbb{C})$ is the simply-connected covering group of G_{i1}; we may choose the covering map to be such that $\hat{T} \cap \hat{G}_{i1}$ is the image of the diagonal matrices and $\sigma_F^{n_i}$ corresponds to the automorphism

$$A \longrightarrow \begin{pmatrix} 0 & 0 & 1 \\ 0 & -1 & 0 \\ 1 & 0 & 0 \end{pmatrix} {}^tA^{-1} \begin{pmatrix} 0 & 0 & 1 \\ 0 & -1 & 0 \\ 1 & 0 & 0 \end{pmatrix}$$

of $SL(3,\mathbb{C})$. We may take w_1 to be the image of

$$\begin{pmatrix} 0 & 0 & 1 \\ 0 & -1 & 0 \\ 1 & 0 & 0 \end{pmatrix} .$$

Then $\sigma_F^{n_i}(w_1) = w_1$.

μ_i acts on $\hat{V}$ as the reflection in the hyperplane perpendicular to β_i . Thus $\mu_1, \ldots, \mu_m$ generate Ω^o . If ω belongs to Ω^o choose w in the normalizer of $\hat{T}$ whose image in Ω is ω . The image of $w\sigma_F(w^{-1})$ in $\hat{T}/\Psi(W_{\mathbb{C}})$ depends only on ω . Call it δ_ω . Then

$$\delta_{\omega_1\omega_2} = w_1w_2\sigma_F(w_2^{-1}w_1^{-1}) = w_1(w_2\sigma_F(w_2^{-1}))w_1^{-1}(w_1\sigma_F(w_1^{-1})) = \omega_1(\delta_{\omega_1})\delta_{\omega_1} .$$

Since δ_ω is 1 on a set of generators this relation shows that it is identically 1 .

Returning to the original problem we show first that if $\chi_{t_1} = \chi_{t_2}$ there is an ω in Ω^o so that $\omega(t_1) = t_2$. Then if w lies in the normalizer of $\hat{T}$ in $\hat{G}$ and its image in Ω is ω we will have $w(t_1\sigma_F)w^{-1} = t_2w\sigma_F(w^{-1})\sigma_F$. Since $w\sigma_F(w^{-1})$ lies in $\Psi(W_{\mathbb{C}})$ the element on the right is conjugate to $t_2\sigma_F$.

If t belongs to $\hat{T}$ let χ_t also denote the homomorphism

$$\Sigma\ \hat{f}(\lambda)\lambda \longrightarrow \Sigma\ \hat{f}(\lambda)\lambda(t)$$

of $\Lambda(\hat{M})$ into $\mathbb{C}$. If there were no ω so that $\omega(t_1) = t_2$ there would be an $\hat{f}$ in $\Lambda(\hat{M})$ so that

$$\chi_{t_2}(\hat{f}) \neq \chi_{\omega(t_1)}(\hat{f})$$

for all ω in Ω^o . Let

$$\prod(X-\omega(\hat{f})) = \sum_{k=o}^{n} \hat{f}_k X^k .$$

Each $\hat{f}_k$ belongs to $\Lambda^o(\hat{M})$. Applying χ_{t_1} and χ_{t_2} we find that

$$\prod_\omega (X-\chi_{\omega(t_1)}(\hat{f})) = \sum_{k=o}^n \chi_{t_1}(\hat{f}_k)X^k = \sum_{k=o}^n \chi_{t_2}(\hat{f}_k)X^k = \prod_\omega (X-\chi_{\omega(t_2)}(\hat{f})) .$$

The polynomial on the right has $\chi_{t_2}(\hat{f})$ as a root but that on the left does not. This is a contradiction.

If $t_1\sigma_F$ and $t_2\sigma_F$ are conjugate then for every representation ρ of $\hat{G}_F$

$$\text{trace}\ \rho(t_1\sigma_F) = \text{trace}\ \rho(t_2\sigma_F) .$$

Let ρ act on X and if λ belongs to $\hat{M}$ let t_λ be the trace of $\rho(\sigma_F)$ on

$$X_\lambda = \{x \in X \mid \rho(t)x = \lambda(t)x \quad \text{for all } t \text{ in } \hat{T}\} .$$

If t belongs to $\Psi(W_{\mathbb{C}})$ then $\lambda(t) = 1$. If ω belongs to Ω^o and w in the normalizer of T has image ω in Ω then $X_{\omega\lambda} = \rho(w)X_\lambda$. Then $t_{\omega\lambda}$ is the trace of $w^{-1}\sigma_F w = w^{-1}\sigma_F(w)\sigma_F$ on X_λ. Since $\lambda(w^{-1}\sigma_F(w)) = 1$ we have $t_{\omega\lambda} = t_\lambda$ and

$$\text{trace}\ \rho(t\sigma_F) = \Sigma_{\lambda\in\hat{C}}\, t_\lambda\, (\Sigma_{\mu\in S(\lambda)}\, \mu(t))$$

if $S(\lambda)$ is the orbit of λ. If

$$\hat{f}_\rho = \Sigma_{\lambda\in\hat{C}}\, t_\lambda\, \Sigma_{\mu\in S(\lambda)}\mu$$

then $\hat{f}_\rho$ belongs to $\Lambda^o(\hat{M})$ and

$$\text{trace}\ \rho(t\sigma_F) = \chi_t(\hat{f}_\rho) .$$

All we need do is show that the elements $\hat{f}_\rho$ generate $\Lambda^0(\hat{M})$ as a vector space. This is an easy induction argument because every λ in $\hat{C}$ is the highest weight of a representation of $\hat{G}_F$ whose restriction to $\hat{G}$ is irreducible.

5. If t belongs to $\hat{T}$ there is a unique function ϕ_t on G_F which satisfies $\phi_t(ug) = \phi_t(gu) = \phi_t(g)$ for all u in U_F and all g in G_F so that

$$\chi_t(f) = \int_{G_F} \phi_t(g)f(g)dg$$

for all f in $C_c(G_F,U_F)$. A formula for ϕ_t, valid under very general assumptions, has been found by I. G. MacDonald. However, because of the present state of reduction theory, his assumptions do not cover the cases in which we are interested. I am going to assume that the obvious generalization of his theorem is valid. In stating it we may as well suppose that t belongs to $\Psi(V_{\mathbb{C}})$.

Let $\hat{N}$ be the unipotent radical of $\hat{B}$, let $\hat{\mathfrak{n}}$ be its Lie algebra, and let τ be the representation of $\hat{T} \times \mathfrak{G}(K/F)$ on $\hat{\mathfrak{n}}$. If t belongs to $\Psi(V_{\mathbb{C}})$ consider the function θ_t on $\hat{M}$ defined by

$$\theta_t(\lambda) = c|\pi_F|^{-\langle\rho,\lambda\rangle} \sum_{\omega\in\Omega^0} \frac{\det(I - |\pi_F|\tau^{-1}(\omega(t)\sigma_F))}{\det(I - \tau^{-1}(\omega(t)\sigma_F))} \lambda^{-1}(\omega(t)) .$$

If $n(\hat{\beta})$ is the number of positive roots projecting onto $\hat{\beta}$ in $\hat{Q}$

$$c = \prod_{\beta\in Q} \left\{ \frac{1 - |\pi_F|^{n(\hat{\beta})\langle\rho,\hat{\beta}\rangle}}{1 - |\pi_F|^{n(\hat{\beta})(\langle\rho,\hat{\beta}\rangle+1)}} \right\} .$$

As it stands $\theta_t(\lambda)$ makes sense only when none of the eigenvalues of $\tau(\omega(t)\sigma_F)$ are 1 for any ω in Ω^o. However using the results of Kostant [8] we can write it in a form which makes sense for all t. Let $\hat{\rho}$ be one-half the sum of the positive coroots. $\hat{\rho}$ belongs to $\hat{V}$. If λ belongs to $\hat{M}$ and $\lambda+\hat{\rho}$ is non-singular, that is $(\lambda+\hat{\rho}, \hat{\beta}) \neq 0$ for all $\hat{\beta}$ in $\hat{Q}$, let ω in Ω^o take $\lambda+\hat{\rho}$ to $\hat{C}$ and let χ_λ be sgn ω times the character of the representation of $\hat{G}_F$ with highest weight $\omega(\lambda+\hat{\rho})-\hat{\rho}$. If $\lambda+\hat{\rho}$ is singular let $\chi_\lambda \equiv 0$. If

$$\det(I-|\pi_F|\tau^{-1}(t\sigma_F)) = \Sigma_{\mu\in\hat{M}}\, b_\mu \mu(t)$$

then

$$\theta_t(\lambda) = c\,|\pi_F|^{-\langle\rho,\lambda\rangle}\, \Sigma_{\mu\in\hat{M}}\, b_\mu\, \chi_{\mu-\lambda}((t\sigma_F)) .$$

Clearly b_μ is 0 unless

$$\mu = -\Sigma_{\hat{\alpha}\in S}\,\hat{\alpha}$$

where S is a subset of the set of positive coroots invariant under $\mathfrak{G}(K/F)$. If U is the collection of such μ then $\{\hat{\rho}+\mu \mid \mu\in M\}$ is invariant under Ω^o. Suppose $\hat{\rho}+\mu$ is non-singular and belongs to $\hat{C}$. Since $\langle\alpha_i, \hat{\rho}\rangle = 1$ and $\langle\alpha_i,\mu\rangle$ is integral, for $1 \leq i \leq \ell$, μ itself must belong to $\hat{C}$. This can only happen if μ is 0. Thus if $b_\mu \neq 0$ either $\hat{\rho}+\mu$ is singular or $\hat{\rho}+\mu$ belongs to the orbit of $\hat{\rho}$ and $\chi_\mu(g) \equiv \pm 1$ on $\hat{G}_F$. As a consequence $\theta_t(0)$ is independent of t. Choose t_o so that $\hat{\beta}_i(t_o) = |\pi_F|^{-\langle\rho,\hat{\beta}_i\rangle}$ for $1 \leq i \leq m$. The eigenvalues of $\tau(\omega(t_o)\sigma_F)$ are the numbers $\zeta|\pi_F|^{-\langle\rho,\omega^{-1}\hat{\beta}\rangle}$ where $\hat{\beta}$ belongs to Q and ζ is an $n(\hat{\beta})$<u>th</u> root of unity. If $\omega \neq 1$ there is a $\hat{\beta}_i$ so that $\omega^{-1}\hat{\beta} = -\hat{\beta}_i$ for some β in Q. Then

$\langle\rho, \omega^{-1}\hat{\beta}\rangle = -\langle\rho,\hat{\beta}_i\rangle = -1$ and $\tau\,(\omega(t_o)\sigma_F)$ has $|\pi_F|$ as an eigenvalue. Thus

$$\theta_{t_o}(0) = c\,\frac{\det\,(I - |\pi_f|\tau^{-1}(t_o\sigma_F))}{\det\,(I - \tau^{-1}(t_o\sigma_F))} = 1\ .$$

We are going to assume that if t belongs to $\Psi(V_{\mathbb{C}})$, a belongs to T_F, and $\lambda = v(a)$, then

$$\phi_t(a) = \theta_t(\lambda)\ .$$

If

$$|\chi_t(f)| \leq \int_{G_F} |f(g)|dg$$

for all f in $C_c(G_F, U_F)$ then ϕ_t is bounded. I want to show that if ϕ_t is bounded, λ belongs to $\hat{L}$, $\overline{\lambda}$ in $\hat{D}$ belongs to the orbit of λ under Ω, and t lies in $\Psi(V_{\mathbb{C}})$. Then

$$|\lambda(t)| \leq |\pi_F|^{-\langle\rho,\overline{\lambda}\rangle}\ .$$

Let $t = \Psi(v)$. v is not determined by t but $\mathrm{Re}\ v$ is and

$$|\lambda(t)| = |\pi_F|^{-\mathrm{Re}\langle v,\lambda\rangle}\ .$$

We will show that if ϕ_t is bounded then $\mathrm{Re}\ \langle v,\lambda\rangle \leq \langle\rho,\overline{\lambda}\rangle$ for all λ in $\hat{E}_{|\mathbb{R}}$. If ω belongs to Ω^o and $\mathrm{Re}\ \omega v$ lies in $\hat{C}$ then $\mathrm{Re}\ \langle\omega v,\omega\lambda\rangle = \mathrm{Re}\ \langle v,\lambda\rangle$. With no loss of generality we may suppose that v lies in C, the analogue of $\hat{C}$. Then, as is well-known,

$$\mathrm{Re}\ \langle v,\lambda\rangle \leq \mathrm{Re}\ \langle v,\overline{\lambda}\rangle$$

and we may as well assume that $\lambda = \bar{\lambda}$. We want to show that $\mathrm{Re}\ \langle v,\lambda\rangle \leq \langle \rho,\lambda\rangle$ for all λ in $\hat{D}$. Since ρ and v both belong to $V_{\mathbb{C}}$ it is sufficient to verify it for λ in $\hat{C}$. The set of λ in $\hat{C}$ for which it is true is closed, convex, and positively homogeneous. Therefore if it contains $\hat{M} \cap \mathrm{Interior}\ \hat{C}$ it is $\hat{C}$.

Let S be the set of simple coroots $\hat{\alpha}$ for which $\mathrm{Re}\ \langle v,\hat{\alpha}\rangle = 0$. Let Σ_o be the positive coroots which are linear combinations of the elements of S and let Σ_+ be the other positive coroots. If $\hat{\mathfrak{n}}_o$ is the span of the root vectors associated to the coroots in Σ_o and $\hat{\mathfrak{n}}_+$ is the span of the root vectors associated to the coroots in Σ_+ then τ breaks up into the direct sum of a representation τ_o on $\hat{\mathfrak{n}}_o$ and a representation τ_+ on $\hat{\mathfrak{n}}_+$. Let $\hat{H}$ be the analytic subgroup of $\hat{G}_F$ whose Lie algebra is generated by the root vectors associated to the coroots of Σ_o and their negatives and let $\mathbb{H}^o$ be the subgroup of Ω^o consisting of those elements with representatives in $\hat{H}$. If ω belongs to Ω^o and $\mathrm{Re}\,\omega v = \mathrm{Re}\ v$ then ω belongs to $\mathbb{H}^o$. If $\mathrm{Re}\,\omega v \neq \mathrm{Re}\,v$ then $\mathrm{Re}\langle \omega v,\lambda\rangle < \mathrm{Re}\langle v,\lambda\rangle$ for λ in $\hat{M} \cap \mathrm{Interior}\ \hat{C}$. Write $\lambda = \lambda_1 + \lambda_2$ where λ_1 is a linear combination of the coroots in S and λ_2 is orthogonal to these roots. If $s = \Psi(u)$ with u in $V_{\mathbb{C}}$ consider

$$\theta_s'(\lambda) =$$

$$c|\pi_F|^{\langle u-\rho,\lambda_2\rangle} \frac{\det(I-|\pi_F|\tau_+^{-1}(s\sigma_F))}{\det(I-\tau_+^{-1}(s\sigma_F))} \left\{ \sum_{\mathbb{H}^o} \frac{\det(I-|\pi_F|\tau_o^{-1}(s\sigma_F))}{\det(I-\tau_o^{-1}(s\sigma_F))} |\pi_F|^{\langle \omega u-\rho,\lambda_1\rangle} \right\}.$$

The function θ_s' is not necessarily defined for all s . However the preceding discussion, applied to $\hat{H}$ rather than $\hat{G}$, shows that it is defined at t and that $\theta_t'(0) \neq 0$. A simple application of ℓ'Hospital's rule shows that, as a function of λ, θ_t' is the product of $|\pi_F|^{\langle v-\rho,\lambda\rangle}$ and a linear combination of products of polynomials and purely imaginary exponentials in λ_1. Thus it does not vanish identically in any open cone.

Set $\theta_t'' = \theta_t - \theta_t'$. θ_t'' is a linear combination of products of polynomials in λ and an exponential $|\pi_F|^{\langle \omega v-\rho,\lambda\rangle}$ with $\mathrm{Re}\,\omega v \neq \mathrm{Re}\, v$. Thus if λ belongs to the interior of $\hat{C}$

$$\lim_{n\to-\infty} |\pi_F|^{\langle \rho-v,n\lambda\rangle} \theta_t''(n\lambda) = 0$$

and

$$\lim_{n\to-\infty} |\pi_F|^{\langle \rho-v,n\lambda\rangle} \theta_t(n\lambda) = \lim_{n\to-\infty} |\pi_F|^{\langle \rho-v,n\lambda\rangle} \theta_t'(n\lambda) .$$

If $\langle\rho,\lambda\rangle$ is less than $\mathrm{Re}\langle v,\lambda\rangle$ for some λ in $\hat{C}$ then $\langle\rho,\lambda\rangle$ is less than $\mathrm{Re}\langle v,\lambda\rangle$ for a λ in $\hat{C}$ for which $\theta_t'(n\lambda)$ does not vanish identically as a function of n. Since ϕ_t is bounded

$$\lim_{n\to-\infty} |\pi_F|^{\langle \rho-v,n\lambda\rangle} \theta_t'(n\lambda) = 0 .$$

But $|\pi_F|^{\langle\rho-v,n\lambda\rangle}\theta_t'(n\lambda)$ is a function of the form

$$\textstyle\sum_{k=o}^{q} \varphi_k(n) n^k$$

where $\varphi_k(n)$ is a linear combination of purely imaginary exponentials e^{ixn}. It is easy to see that it cannot approach 0 as n approaches $-\infty$.

6. Suppose G is a group defined over the global field F. There is a quasi-split group G' over F and an isomorphism $\varphi: G \longrightarrow G'$ defined over a Galois extension K of F so that, for every σ in $\mathfrak{G}(K/F)$, $a_\sigma = \varphi^\sigma\varphi^{-1}$

is an inner automorphism of G'. We assume that there is a lattice $\mathfrak{g}_{O_F}$ over O_F in the Lie algebra of G' so that $O_K\mathfrak{g}_{O_F}$ is a Chevalley lattice.

If $\mathfrak{q}$ is a finite prime of K and $\mathfrak{P}$ is a prime of K dividing $\mathfrak{q}$ the group G over $F_{\mathfrak{q}}$ is obtained from G' by twisting by the restriction $\bar{a}$ of the cocycle $\{a_\sigma\}$ to $\mathfrak{G}(K_{\mathfrak{P}}/F_{\mathfrak{q}})$. Let $\bar{G}'$ be the adjoint group of G'. If $\bar{U}'_{K_{\mathfrak{P}}}$ is the stabilizer of the lattice $O_{K_{\mathfrak{P}}}\mathfrak{g}_{O_F}$ then, for almost all $\mathfrak{q}$, $\bar{a}$ takes values in $\bar{U}'_{K_{\mathfrak{P}}}$. If $K_{\mathfrak{P}}/F_{\mathfrak{q}}$ is also unramified then G is quasi-split over $F_{\mathfrak{q}}$ because $H^1(\mathfrak{G}(K_{\mathfrak{P}}/F_{\mathfrak{q}}), \bar{U}'_{K_{\mathfrak{q}}}) = \{1\}$. Let S be the set of those $\mathfrak{q}$, unramified in K, for which $\bar{a}$ takes values in $\bar{U}'_{K_{\mathfrak{P}}}$. Let G act on a vector space X over F and let X_{O_F} be a lattice in X_F. Let $U_{F_{\mathfrak{q}}}$ be the stabilizer of $O_{F_{\mathfrak{q}}}X_{O_F}$ in $G_{F_{\mathfrak{q}}}$ and let $U'_{F_{\mathfrak{q}}}$ be the stabilizer of $O_{F_{\mathfrak{q}}}\mathfrak{g}_{O_F}$ in $G'_{F_{\mathfrak{q}}}$. Then $\varphi(U_{F_{\mathfrak{q}}}) = U'_{F_{\mathfrak{q}}}$ for almost all $\mathfrak{q}$. If $\mathfrak{q}$ is also in S choose u in $\bar{U}'_{F_{\mathfrak{q}}}$ so that $\varphi^\sigma\varphi^{-1} = \mathrm{Ad}\, u^\sigma u^{-1}$ for all σ in $\mathfrak{G}(K_{\mathfrak{P}}/F_{\mathfrak{q}})$. Then $\varphi^{-1}\mathrm{Ad}u$ is defined over F and $\varphi^{-1}\mathrm{Ad}u(U'_{F_{\mathfrak{q}}}) = U_{F_{\mathfrak{q}}}$. Consequently $U_{F_{\mathfrak{q}}}$ is one of the compact subgroups of the fourth paragraph.

To show that almost all $\mathfrak{q}$ are unramified all we need do is observe that if π occurs in $L^2(G_F\backslash G_{A(F)})$, whatever the precise meaning of this is to be, and $\pi = \prod_{\mathfrak{q}}\otimes\pi_{\mathfrak{q}}$ then for almost all $\mathfrak{q}$ the restriction of $\pi_{\mathfrak{q}}$ to $U_{F_{\mathfrak{q}}}$ contains the trivial representation.

If $\mathfrak{p}$ is unramified let the homomorphism of $C_c(G_{F_\mathfrak{p}}, U_{F_\mathfrak{p}})$ associated to $\pi_\mathfrak{p}$ be $\chi_{t_\mathfrak{p}}$. To show that the product of the local L-functions converges in a half plane it would be enough to show that there is a positive constant a so that for all unramified $\mathfrak{p}$ every eigenvalue of $\rho(t_\mathfrak{p}\sigma_{F_\mathfrak{p}})$ is bounded by $|\pi_\mathfrak{p}|^{-a}$. We may suppose that $\sigma_{F_\mathfrak{p}}(t_\mathfrak{p}) = t_\mathfrak{p}$. If $n = [K:F]$ then $(t_\mathfrak{p}\sigma_{F_\mathfrak{p}})^n = t_\mathfrak{p}^n$ so that we need only show that the eigenvalues of $\rho(t_\mathfrak{p})$ are bounded by $|\pi_\mathfrak{p}|^{-a}$. This we did in the previous paragraph.

7. Once the definitions are made we can begin to pose questions. My hope is that all these questions have affirmative answers. The first question is the one initially posed.

<u>Question 1</u>. <u>Is it possible to define the local L-functions</u> $L(s,\rho,\pi)$ <u>and the local factors</u> $\varepsilon(s,\rho,\pi,\Psi_F)$ <u>at the ramified primes so that if</u> F <u>is a global field</u>, $\pi = \prod \otimes \pi_\mathfrak{p}$, <u>and</u>

$$L(s,\rho,\pi) = \prod_\mathfrak{p} L(s,\rho_\mathfrak{p},\pi_\mathfrak{p})$$

<u>then</u> $L(s,\rho,\pi)$ <u>is meromorphic in the entire complex plane with only a finite number of poles and satisfies the functional equation</u>

$$L(s,\rho,\pi) = \varepsilon(s,\rho,\pi)L(1-s,\tilde{\rho},\pi)$$

<u>with</u>

$$\varepsilon(s,\rho,\pi) = \prod_\mathfrak{p} \varepsilon(s,\rho_\mathfrak{p},\pi_\mathfrak{p},\Psi_{F_\mathfrak{p}}) .$$

The theory of Eisenstein series can be used [9] to give some novel instances in which this question has, in part, an affirmative answer. However that theory does not suggest any method of attacking the general problem. If $G = GL(n)$ then $\hat{G}_F = GL(n,\mathbb{C})$. The work of Godement and earlier writers allows one to hope that the methods of Hecke and Tate can, once the representation theory of the general linear group over a local field is understood, be used to answer the first question when $G = GL(n)$ and ρ is the standard representation of $GL(n,\mathbb{C})$. The idea which led Artin to the general reciprocity law suggests that we try to answer it in general by answering a further series of questions. For the sake of precision, but not clarity, I write them down in an order opposite to that in which they suggest themselves. If G is defined over the local field F let $\Omega(G_F)$ be the set of equivalence classes of irreducible unitary representations of G_F .

Question 2. *Suppose G and G' are defined over the local field F, G is quasi-split and G' is obtained from G by an inner twisting. Then $\hat{G}_F = \hat{G}'_F$. Is there a correspondence R whose domain is $\Omega(G'_F)$ and whose range is contained in $\Omega(G_F)$ so that if $\pi = R(\pi')$ then $L(s,\rho,\pi) = L(s,\rho,\pi')$ for every representation ρ of $\hat{G}_F$?*

Notice that R is not required to be a function. I do not know whether or not to expect that

$$\varepsilon(s,\rho,\pi,\Psi_F) = \varepsilon(s,\rho,\pi',\Psi_F) .$$

One should, but I have not yet done so, look carefully at this question when F is the field of real numbers. For this one will of course need the work of Harish-Chandra.

Supposing that the second question has an affirmative answer one can formulate a global version.

Question 3.[*] *Suppose that* G *and* G' *are defined over the global field* F, G *is quasi-split, and* G' *is obtained from* G *by an inner twisting. Suppose* $\pi' = \prod_{\mathfrak{p}} \otimes \pi'_{\mathfrak{p}}$ *occurs in* $L^2(G'_F \backslash G'_{\mathbb{A}(F)})$. *Choose for each* $\mathfrak{p}$ *a representation* $\pi_{\mathfrak{p}}$ *of* $G_{F_{\mathfrak{p}}}$ *so that* $\pi_{\mathfrak{p}} = R(\pi'_{\mathfrak{p}})$. *Does* $\pi = \prod_{\mathfrak{p}} \otimes \pi_{\mathfrak{p}}$ *occur in* $L^2(G_F \backslash G_{\mathbb{A}(F)})$?

Affirmative evidence is contained in papers of Eichler [3] and Shimizu [16] when $G = GL(2)$ and G' is the group of invertible elements in a quaternion algebra. Jacquet [6], whose work is not yet complete, is obtaining very general results for these groups.

Question 4. Suppose G *and* G' *are two quasi-split groups over the local field* F. *Let* G *split over* K *and let* G' *split over* K' *with* $K \subseteq K'$. *Let* Ψ *be the natural map* $\mathcal{W}(K'/F) \longrightarrow \mathcal{W}(K/F)$. *Suppose* φ *is a complex analytic homomorphism from* $\hat{G}'_{K'/F}$ *to* $\hat{G}_{K/F}$ *which makes*

$$\begin{array}{ccc}
\hat{G}'_{K'/F} & \longrightarrow & \mathcal{W}(K'/F) \\
\Big\downarrow \varphi & & \Big\downarrow \Psi \\
\hat{G}_{K/F} & \longrightarrow & \mathcal{W}(K/F)
\end{array}$$

commutative. Is there a correspondence R_{φ} *with domain* $\Omega(G'_F)$ *whose range is contained in* $\Omega(G_F)$ *so that if* $\pi = R_{\varphi}\pi'$ *then, for every representation*

---

[*] The question, in this crude form, does not always have an affirmative answer (cf. [6]). The proper question is certainly more subtle but not basically different.

ρ *of* $\hat{G}_F$ *and every non-trivial additive character* ψ_F, $L(s,\rho,\pi) = L(s,\rho\circ\varphi,\pi')$ *and* $\varepsilon(s,\rho,\pi,\psi_F) = \varepsilon(s,\rho\circ\varphi,\pi',\psi_F)$?

R_φ should of course be functorial and, in an unramified situation, if π' is associated to the conjugacy class $t' \times \sigma'_F$ then π should be associated to $\varphi(t' \times \sigma'_F)$. I have not yet had a chance to look carefully at this question when F is the field of real numbers.

The question has a global form.

Question 5. Suppose G *and* G' *are two quasi-split groups over the global field* F . *Let* G *split over* K *and let* G' *split over* K' *with* $K \subseteq K'$. *Suppose* φ *is a complex analytic homomorphism from* $\hat{G}'_{K'/F}$ *to* $\hat{G}_{K/F}$ *which makes*

$$\begin{array}{ccc}
\hat{G}'_{K'/F} & \longrightarrow & \mathfrak{G}(K'/F) \\
\Big\downarrow \varphi & & \Big\downarrow \\
\hat{G}_{K/F} & \longrightarrow & \mathfrak{G}(K/F)
\end{array}$$

commutative. If $\mathfrak{P}'$ *is a prime of* K' *let* $\mathfrak{P} = \mathfrak{P}' \cap K$ *and let* $\mathfrak{p} = \mathfrak{P}' \cap F$. φ *determines a homomorphism* $\varphi_{\mathfrak{p}}: \hat{G}'_{K'_{\mathfrak{P}'}/F_{\mathfrak{p}}} \longrightarrow \hat{G}_{K_{\mathfrak{P}}/F_{\mathfrak{p}}}$ *which makes*

$$\begin{array}{ccc}
\hat{G}'_{K'_{\mathfrak{P}'}/F_{\mathfrak{p}}} & \longrightarrow & \mathfrak{G}(K'_{\mathfrak{P}'}/F_{\mathfrak{p}}) \\
\Big\downarrow & & \Big\downarrow \\
\hat{G}_{K_{\mathfrak{P}}/F_{\mathfrak{p}}} & \longrightarrow & \mathfrak{G}(K_{\mathfrak{P}}/F_{\mathfrak{p}})
\end{array}$$

commutative. If $\pi' = \prod \otimes \pi'_{\mathfrak{p}}$ *occurs in* $L^2(G'_F \backslash G'_{A(F)})$ *choose for each*

$\mathfrak{p}$ *a* $\pi_{\mathfrak{p}} = R_{\varphi_{\mathfrak{p}}}(\pi'_{\mathfrak{p}})$. *If* $\pi = \prod_{\mathfrak{p}} \otimes \pi_{\mathfrak{p}}$ *does* π *occur in* $L^2(G_F \backslash G_{\mathbb{A}(F)})$?

An affirmative answer to the third and fifth questions would allow us to solve the first question by examining automorphic forms on the general linear groups.

It is probably worthwhile to point out the difficulty of the fifth question by giving some examples. Take $G' = \{1\}$ $G = GL(1)$, K' any Galois extension of F, and $K = F$. The assertion that, in this case, the last two questions have affirmative answers is the Artin reciprocity law.

Suppose G is quasi-split and $G' = T$. We may identify $\hat{G}'_F$ with $\hat{T} \times \mathfrak{G}(K/F)$ which is contained in $\hat{G}_F$. Thus we take $K' = K$. Let φ be the imbedding. In this case π' is a character of $G'_F \backslash G'_{\mathbb{A}(F)}$. The fourth question is, with certain reservations, answered affirmatively by the theory of induced representations. The fifth question is, with similar reservations, answered by the theory of Eisenstein series. The reservations are not important. I only want to point out that the theory of Eisenstein series is a prerequisite to the solution of these problems. With G as before take $G'' = \{1\}$ and $K'' = K$ so that $\hat{G}''_F = \mathfrak{G}(K/F)$. Let Ψ take σ in $\mathfrak{G}(K/F)$ to σ in $\hat{G}_F$. There is only one choice for π''. The associated space of automorphic forms on $G_F \backslash F_{\mathbb{A}(F)}$ should be the space of automorphic forms associated to the trivial character of $G'_F \backslash G'_{\mathbb{A}(F)}$. For this character all the reservations apply. I point out that the space associated to π'' is not the obvious one. It is not the space of constant functions. To prove its existence will require the theory of Eisenstein series.

Take $G = GL(2)$ and let G' be the multiplicative group of a separable quadratic extension K' of F. Take $K = F$. Then $\hat{G}'_F$ is a semi-direct product $(\mathbb{C}^* \times \mathbb{C}^*) \times \mathfrak{G}(K'/F)$. If σ is the non-trivial element of $\mathfrak{G}(K'/F)$ then $\sigma((t_1,t_2)) = (t_2,t_1)$. Let φ be defined by

$$\varphi : (t_1,t_2) \longrightarrow \begin{pmatrix} t_1 & 0 \\ 0 & t_2 \end{pmatrix}$$

$$\varphi : \sigma \longrightarrow \begin{pmatrix} 0 & 1 \\ 1 & 0 \end{pmatrix}$$

The existence of R_φ in the local case is a known fact (see, for example, [6]) in the theory of representations of GL(2,F) . An affirmative answer to the fifth question can be given by means of the Hecke theory [6] and by other means [15].

Let E be a separable extension of F and let G be the group over F obtained from GL(2) over E by restriction of scalars. Let G' be GL(2) over F and let $K' = K$ be any Galois extension containing E . Let X be the homogeneous space $\mathfrak{G}(K/E)\backslash\mathfrak{G}(K/F)$. Then $\hat{G}_F$ is the semi-direct product of $\prod_{x \in X} GL(2,\mathbb{C})$ and $\mathfrak{G}(K/F)$. If σ belongs to $\mathfrak{G}(K/F)$ then

$$\sigma(\prod_{x \in X} A_x)\sigma^{-1} = \prod_{x \in X} B_x$$

with $B_x = A_{x\sigma}$. Define φ by

$$\varphi(A \times \sigma) = (\prod_{x \in X} A) \times \sigma \quad .$$

Although not much is known about the fifth question in this case the paper [2] of Doi and Naganuma is encouraging.

Suppose G and K are given. Let $G' = \{1\}$ and let K' by any Galois extension of F containing K. If F is a local field the fourth question asks that to every homomorphism φ of $\mathfrak{G}(K'/F)$ into $\hat{G}_F$ which makes

$$\begin{array}{ccc} \mathfrak{G}(K'/F) & \xrightarrow{\ \varphi\ } & \hat{G}_F \\ & \searrow & \downarrow \\ & & \mathfrak{G}(K/F) \end{array}$$

commutative there be associated at least one irreducible unitary representation of G_F. If F is global the fifth question asks that to φ there be associated a representation of $G_{\mathbb{A}(F)}$ occurring in $L^2(G_F \backslash G_{\mathbb{A}(F)})$.

The L-functions we have introduced have been so defined that they include the Artin L-functions. However Weil [17] has generalized the notion of an Artin L-function. The preceding observations suggest a relation between the generalized Artin L-functions and the L-functions of this paper. Weil's definition requires the introduction of some locally compact groups - the Weil groups. If F is a local field let C_F be the multiplicative group of F. If F is a global field let C_F be the idèle class group. If K is a Galois extension of F the Weil group $W_{K/F}$ is an extension

$$1 \longrightarrow C_K \longrightarrow W_{K/F} \longrightarrow \mathfrak{G}(K/F) \longrightarrow 1$$

of $\mathfrak{G}(K/F)$ by C_K. There is a canonical homomorphism $\tau_{K/F}$ of $W_{K/F}$ onto C_F. If F is a global field, $\mathfrak{P}$ a prime of K, and $\mathfrak{p} = F \cap \mathfrak{P}$ there is a homomorphism $\alpha_{\mathfrak{p}} : W_{K_{\mathfrak{P}}/F_{\mathfrak{p}}} \longrightarrow W_{K/F}$. $\alpha_{\mathfrak{p}}$ is determined up to an inner automorphism. If σ is a representation of $W_{K/F}$ the class of $\sigma_{\mathfrak{p}} = \sigma \circ \alpha_{\mathfrak{p}}$

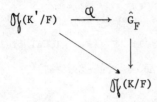

is independent of $\alpha_{\mathfrak{p}}$. By a representation σ of $W_{K/F}$ we understand a finite dimensional complex representation such that $\sigma(w)$ is semi-simple for all w in $W_{K/F}$.

If F is a local field and ψ_F a non-trivial additive character of F then for any representation σ of $W_{K/F}$ we can define (cf. [11]) a local L-function $L(s,\sigma)$ and a factor $\varepsilon(s,\sigma,\psi_F)$. If F is a global field and σ is a representation of $W_{K/F}$ the associated L-function is

$$L(s,\sigma) = \prod_{\mathfrak{p}} L(s,\sigma_{\mathfrak{p}}) .$$

The product is taken over all primes including the archimedean ones. If ψ_F is a non-trivial character of $F\backslash A(F)$ then $\varepsilon(s,\sigma_{\mathfrak{p}},\psi_{F_{\mathfrak{p}}})$ is 1 for almost all $\mathfrak{p}$,

$$\varepsilon(s,\sigma) = \prod_{\mathfrak{p}} \varepsilon(s,\sigma_{\mathfrak{p}},\psi_{F_{\mathfrak{p}}})$$

is independent of ψ_F, and

$$L(s,\sigma) = \varepsilon(s,\sigma)L(1-s,\tilde{\sigma})$$

if $\tilde{\sigma}$ is contragredient to σ.

Question 6. Suppose G is quasi-split over the local field F and splits over the Galois extension K. Let $\hat{U}_F$ be a maximal compact subgroup of $\hat{G}_F$. Let K' be a Galois extension of F which contains K and let φ be a homomorphism of $W_{K'/F}$ into $\hat{U}_F$ which makes

$$\begin{array}{ccc} W_{K'/F} & \longrightarrow & \mathfrak{G}(K'/F) \\ \downarrow \varphi & & \downarrow \\ \hat{U}_F & \longrightarrow & \mathfrak{G}(K/F) \end{array}$$

commutative. Is there an irreducible unitary representation $\pi(\varphi)$ *of* G_F *so that, for every representation* σ *of* $\hat{G}_F$, $L(s,\sigma,\pi(\varphi)) = L(s,\sigma \circ \varphi)$ *and* $\varepsilon(s,\sigma,\pi(\varphi),\psi_F) = \varepsilon(s,\sigma \circ \varphi, \psi_F)$?

Changing φ by an inner automorphism $\hat{U}_F$ will not change $\pi(\varphi)$ or at least not its equivalence class. If F in non-archimedean and K'/F is unramified the composition of v , the valuation on F , and $\tau_{K'/F}$ defines a homomorphism ω of $W_{K/F}$ onto $\mathbb{Z}$. If $u = t \times \sigma_F$ belongs to $\hat{U}_F$ we could define φ by

$$\varphi(w) = u^{\omega(w)} .$$

Then $\pi(\varphi)$ would be the representation associated to the homomorphism χ_t of the Hecke algebra into $\mathbb{C}$.

We can also ask the question globally.

Question 7. Suppose G *is quasi-split over the global field* F *and splits over* K . *Let* K' *be a Galois extension of* F *containing* K *and let* φ *be a homomorphism of* $W_{K'/F}$ *into* $\hat{U}_F$ *which makes*

$$\begin{array}{ccc} W_{K'/F} & \longrightarrow & \mathfrak{G}(K'/F) \\ \downarrow \varphi & & \downarrow \\ \hat{U}_F & \longrightarrow & \mathfrak{G}(K/F) \end{array}$$

commutative. If $\mathfrak{p}'$ *is a prime of* K' *and* $\mathfrak{p} = \mathfrak{p}' \cap F$ *then* $\varphi_{\mathfrak{p}} = \varphi \circ \alpha_{\mathfrak{p}}$ *takes* $W_{K'_{\mathfrak{p}'}/F_{\mathfrak{p}}}$ *into* $\hat{U}_{F_{\mathfrak{p}}}$. *If* $\pi(\varphi) = \prod_{\mathfrak{p}} \otimes \pi(\varphi_{\mathfrak{p}})$ *does* $\pi(\varphi)$ *occur in* $L^2(G_F \backslash G_{\mathbb{A}(F)})$?

Both questions have affirmative answers if G is abelian [10] and the correspondence $\varphi \longleftrightarrow \pi(\varphi)$ is surjective. In this case our L-functions are all generalized Artin L-functions. If $G = GL(2)$ and $K = F$ it appears that the Hecke theory can be used to give an affirmative answer to both questions if it is assumed that certain of the generalized Artin L-functions have the expected analytic properties. If all goes well the details will appear in [6].

I would like very much to end this series of questions with some reasonably precise questions about the relation of the L-functions of this paper to those associated to non-singular algebraic varieties. Unfortunately I am not competent to do so. Since it may be of interest I would like to ask one question about the L-functions associated to elliptic curves. If C is defined over a local field F of characteristic zero I am going to associate to it a representation $\pi(C/F)$ of $GL(2,F)$. If C is defined over a global field F which is also characteristic zero then, for each prime $\mathfrak{p}$, $\pi(C/F_{\mathfrak{p}})$ is defined. Does $\pi = \prod_{\mathfrak{p}} \otimes\, \pi(C/F_{\mathfrak{p}})$ occur in $L^2(GL(2,F)\backslash GL(2,\mathbb{A}(F))$? If so $L(s,\sigma,\pi)$, with σ the standard representation of $GL(2,\mathbb{C})$, whose analytic properties are known [6] will be one of the L-functions associated to the elliptic curve. There are examples on which the question can be tested. I hope to comment on them in [6].

To define $\pi(C/F)$ I use the results of Serre [14]. Suppose that F is non-archimedean and the j-invariant of C is integral. Take any prime ℓ different from the characteristic of the residue field and consider the ℓ-adic representation. There is a finit Galois extension K of F so that if A is the maximal unramified extension of K the ℓ-adic representation can be regarded as a representation of $\mathfrak{G}(A/F)$. There is a homomorphism of $W_{K/F}$ into $\mathfrak{G}(A/F)$. The ℓ-adic representation of $\mathfrak{G}(A/F)$ determines a representation φ of $W_{K/F}$ in $GL(2,R)$ where R is a finitely generated subfield of the ℓ-adic field $\mathbb{Q}_\ell$. Let σ be an isomorphism of R with a subfield of $\mathbb{C}$. Then

$$\Psi : w \longrightarrow |\tau_{K/F}(w)|^{1/2} \varphi^{\sigma}(w)$$

is a representation of $W_{K/F}$ in a maximal compact subgroup of $GL(2,\mathbb{C})$. Let $\pi(C/F)$ be the representation $\pi(\Psi)$ of Question 6. If C has good reduction the class of Ψ is independent of ℓ and σ . I do not know if this is so in general. It does not matter because we do not demand that $\pi(C/F)$ be uniquely determined by C .

If the j-invariant is not integral the ℓ-adic representation can be put in the form

$$\sigma \longrightarrow \begin{pmatrix} \chi_1(\sigma) & * \\ 0 & \chi_2(\sigma) \end{pmatrix}$$

where χ_1 and χ_2 are two representations of the Galois group of the algebraic closure of F in the multiplicative group of $\mathbb{Q}_\ell$. If A is the maximal abelian extension of F then χ_1 and χ_2 may be regarded as representations of $\mathfrak{G}(A/F)$. There is a canonical map of F^* , the multiplicative group of F , into $\mathfrak{G}(A/F)$. χ_1 and χ_2 thus define characters μ_1 and μ_2 of F^* . μ_1 and μ_2 take values in $\mathbb{Q}^*$ and $\mu_1\mu_2(x) = \mu_1\mu_2^{-1}(x) = |x|^{-1}$. In, for example, [6] there is associated to the pair of generalized characters $x \longrightarrow |x|^{1/2} \mu_1(x)$ and

$x \longrightarrow |x|^{1/2}\mu_2(x)$ a unitary representation of $GL(2,F)$, a so-called special representation. This we take as $\pi(C/F)$.

If F is $\mathbb{C}$ take $\pi(C/F)$ to be the representation of $GL(2,\mathbb{C})$ associated to the map

$$z \longrightarrow \begin{pmatrix} \frac{z}{|z|} & 0 \\ 0 & \frac{\bar{z}}{|z|} \end{pmatrix}$$

of $\mathbb{C}^* = W_{\mathbb{C}/\mathbb{C}}$ into $GL(2,\mathbb{C})$ by Question 6. $\mathbb{C}^*$ is of index two in $W_{\mathbb{C}/\mathbb{R}}$. The representation of $W_{\mathbb{C}/\mathbb{R}}$ induced from the character $z \longrightarrow \frac{z}{|z|}$ of $\mathbb{C}^*$ has degree 2. If $F = \mathbb{R}$ let $\pi(C/F)$ be the representation of $GL(2,\mathbb{R})$ associated to the induced representation by Question 6.

8. I would like to finish up with some comments on the relation of the L-functions of this paper to Ramanujan's conjecture and its generalizations. Suppose $\pi = \prod \otimes \pi_{\mathfrak{p}}$ occurs in the space of cusp forms. The most general form of Ramanujan's conjecture would be that for all $\mathfrak{p}$ the character of $\pi_{\mathfrak{p}}$ is a tempered distribution [5]. However neither the notion of a character nor that of a tempered distribution has been defined for non-archimedean fields. A weaker question is whether or not at all unramified non-archimedean primes the conjugacy class in $\hat{G}_F$ associated to $\pi_{\mathfrak{p}}$ meets $\hat{U}_F$ (cf. [13]). If this is so it should be reflected in the behavior of the L-functions.

Suppose, to remove all ramification, that G is a Chevalley group and that $K = F = \mathbb{Q}$. Suppose also that each $\pi_{\mathfrak{p}}$ is unramified. If p is non-archimedean there is associated to π_p a conjugacy class $\{t_p\}$ in $G_{\mathbb{Q}}$.

We may take t_p in $\hat{T}$. The conjecture is that, for all λ in $\hat{L}$,

$$|\lambda(t_p)| = 1 .$$

Since there is no ramification at ∞ one can, as in [9], associate to π_∞ a semi-simple conjugacy class $\{X_\infty\}$ in the Lie algebra of $\hat{G}_{\mathbf{Q}}$. We may take X_∞ in the Lie algebra of $\hat{T}$. The conjecture at ∞ is that, for λ in $\hat{L}$,

$$\operatorname{Re} \lambda(X_\infty) = 0 .$$

If σ is a complex analytic representation of $\hat{G}_{\mathbf{Q}}$ let $m(\lambda)$ be the multiplicity with which λ occurs in σ. Then

$$L(s,\sigma,\pi) = \prod_\lambda \left\{ \pi^{\frac{-(s+\lambda(X_\infty))}{2}} \Gamma\left(\frac{s+\lambda(X_\infty)}{2}\right) \prod_p \frac{1}{1-\frac{\lambda(t_p)}{p^s}} \right\}^{m(\lambda)} .$$

If the conjecture is true $L(s,\sigma,\pi)$ is analytic to the right of $\operatorname{Re} s = 1$ for all σ.

Let F be any non-archimedean local field and G any quasi-split group over F which splits over an unramified extension field. If f belongs to $C_c(G_F,U_F)$ let $f^*(g) = \overline{f}(g^{-1})$. If $\hat{f}$ and $\hat{f}^*$ are the images of f and f^* in $\Lambda^o(M)$ then $\hat{f}^*(\lambda)$ is the complex conjugate of $\hat{f}(-\lambda)$. If t belongs to $\hat{T}$ define t^* by the condition that $\lambda(t^*) = \overline{\lambda(t^{-1})}$ for all λ in $\hat{L}$. The complex conjugate of $\chi_t(f^*)$ is

$$\Sigma\, \hat{f}(-\lambda)\, \overline{\lambda(t)} = \Sigma\, \hat{f}(\lambda)\, \lambda(t^*) = \chi_{t^*}(f) .$$

If χ_t is the homomorphism associated to a unitary representation then $\chi_t(f^*)$ is the complex conjugate of $\chi_t(f)$ for all f so that $t \times \sigma_F$

is conjugate to $t^* \times \sigma_F$ and for any representation ρ of $\hat{G}_F$ the complex conjugate of trace $\rho(t \times \sigma_F)$ is trace $\tilde{\rho}(t \times \sigma_F)$ if $\tilde{\rho}$ is the contragredient of ρ. In the case under consideration when $K = F$ this means that trace $\rho(t_p)$ is the complex conjugate of trace $\tilde{\rho}(t_p)$. A similar argument can be applied at the infinite prime to show that the eigenvalues of $\rho(X_\infty)$ are the complex conjugates of the eigenvalues of $\tilde{\rho}(X_\infty)$.

Suppose $L(s,\sigma,\pi)$ is analytic to the right of $\mathrm{Re}\, s = 1$ for all σ. Since the Γ-function has no zeros

$$\prod_\lambda \left\{ \prod_p \frac{1}{1 - \frac{\lambda(t_p)}{p^s}} \right\}^{m(\lambda)} \tag{C}$$

is also. Let σ be $\rho \otimes \tilde{\rho}$. Then the logarithm of this Dirichlet series is

$$\Sigma_p \Sigma_{n=1}^{\infty} \frac{\mathrm{trace}\ \sigma^n(t_p)}{p^{ns}} .$$

Since

$$\mathrm{trace}\ \sigma^n(t_p) = \mathrm{trace}\ \rho^n(t_p)\ \mathrm{trace}\ \tilde{\rho}^n(t_p) = |\mathrm{trace}\ \rho^n(t_p)|^2$$

the series for the logarithm has positive coefficients. Thus the original series does too. By Landau's theorem it converges absolutely for $\mathrm{Re}\, s > 1$ and so does the series for its logarithm. In particular

$$\det\left(1 - \frac{\sigma(t_p)}{p^s}\right)$$

does not vanish for $\mathrm{Re}\, s > 1$ so that the eigenvalues of $\sigma(t_p)$ are all less than or equal to p in absolute value. If λ is a weight choose ρ

REFERENCES

Bruhat and J. Tits, Groupes algébriques simples sur un corps local, Driebergen Conference on Local Fields, Springer-Verlag, 1967.

Doi and H. Naganuma, On the algebraic curves uniformized by arithmetical automorphic functions, Ann. of Math., vol. 86 (1967).

Eichler, Quadratische Formen und Modulfunktion, Acta, Arith., vol. 4 (1958).

Gantmacher, Canonical representation of automorphisms of a complex semi-simple Lie group, Mat. Sb. vol. 47 (1939).

rish-Chandra, Discrete Series for Semi-Simple Lie Groups II, Acta Math., vol. 116 (1966).

Jacquet and R. P. Langlands, Automorphic Forms on GL(2), in preparation.

. Jacobson, Lie Algebras, Interscience, 1962.

. Kostant, Lie Algebra Cohomology and the Generalized Borel-Weil Theorem, Ann. of Math., vol. 74 (1961).

P. Langlands, Euler Products, Lecture notes, Yale University (1967).

__________, Representations of Abelian Algebraic Groups, Notes, Yale University (1968).

__________, On the Functional Equation of the Artin L-functions, in preparation.

so that $m\lambda$ occurs in ρ . Then $m\lambda(t_p) = \lambda(t_p)^m$ is an e
and $\overline{\lambda(t_p)^m}$ is an eigenvalue of $\tilde{\rho}$ so that $|\lambda(t_p)|^{2m}$ is
of σ and

$$|\lambda(t_p)| \leqslant p^{\frac{1}{2m}}$$

for all m and all λ . Thus $|\lambda(t_p)| \leqslant 1$ for all λ .
$-\lambda$ we see that $|\lambda(t_p)| = 1$ for all λ . Since the funct
(C) cannot vanish for $\mathrm{Re}\, s > 1$ when $\sigma = \rho \otimes \tilde{\rho}$ the funct

$$\prod_{\lambda} \Gamma\left(\frac{s + \lambda(X_\infty)}{2}\right)^{m(\lambda)}$$

must be analytic for $\mathrm{Re}\, s > 1$. This implies that

$$\mathrm{Re}\, \lambda(X_\infty) \geqslant -1$$

if $m(\lambda) > 0$. The same argument as before leads to the co
$\mathrm{Re}\, \lambda(X_\infty) = 0$ for all λ .

Granted the generalizations of Ramanujan's conjecture
about the asymptotic distribution of the conjugacy classes
make no guesses about the answer. In general it is not poss
the eigenvalues of the Hecke operators in an elementary fash
Question 7 cannot be expected to lead by itself to elementar
laws. However when the groups $G_{F_{\mathscr{Y}}}$ at the infinite primes
compact these eigenvalues should have an elementary meaning.
together with some information on the range of the correspon
3 may eventually lead to elementary, but extremely complicat
laws. At the present it is impossible even to speculate.

1. F
2. K
3. M
4. F
5. H
6. H
7. N
8. F
9. R
10. —
11. —

12. I. Satake, Theory of Spherical Functions on Reductive Algebraic Groups over $\mathfrak{p}$-adic Fields, Publ. Math. No. 18, I.H.E.S. (1963).

13. ________, Spherical Functions and Remanujan Conjecture, in Algebraic Groups and Discontinuous Subgroups, Amer. Math. Soc. (1966).

14. J. P. Serre, Groupes de Lie ℓ-adic attachés aux courbes elliptiques, Colloque de Clermont-Ferrand (1964).

15. J. A. Shalika and S. Tanaka, On an Explicit Construction of a Certain Class of Automorphic Forms, Notes, Institute for Advanced Study (1968).

16. H. Shimizu, On Zeta functions of quaternion algebras, Ann. of Math. vol. 81 (1966).

17. A. Weil, Sur la Theorie du Corps de Classes, Jour. Math. Soc. Japan, vol. 3 (1951).

RANDOM LINEAR FUNCTIONALS: SOME RECENT RESULTS

by R. M. Dudley

1. Generalities (see [5] for further details). Given a probability space (Ω,P), let $L^0(\Omega,P)$ be the vector space of all real-valued measurable functions on the set Ω, modulo functions vanishing almost everywhere. Then for $p > 0$ we have $L^p(\Omega,P) = \{f \in L^0(\Omega,P): \int|f|^p\,dP < \infty\}$. For $p \geq 1$, L^p is metrized by its usual norm. For $0 < p < 1$, L^p is metrized by

$$d_p(f,g) = \int|f-g|^p\,dP$$

and is generally not locally convex. On L^0 we put the metric

$$d(f,g) = \int|f-g|/(1+|f-g|)\,dP\,.$$

d metrizes the topology of convergence in probability (measure) and is usually far from being locally convex.

Given a real vector space S, let S^a be the "algebraic dual" vector space of all real linear functionals on S. Let $\mathcal{M}(S^a,S)$ be the smallest σ-algebra of subsets of S^a making each evaluation $f \to f(s)$ measurable, $s \in S$. A probability measure (countably additive) on $\mathcal{M}(S^a,S)$ will be said to define a random linear functional on S.

A linear map L from S into some $L^0(\Omega,P)$ will be called a linear process. It can be shown, using Kolmogorov's existence theorem for stochastic processes, that for any such L there exists a unique probability P_L on $\mathcal{M}(S^a,S)$ such that for any finite sequence $s_1, \ldots, s_n$ in S, the joint probability distribution of $L(s_1), \ldots, L(s_n)$ for P equals that of $f(s_1), \ldots, f(s_n)$ for f distributed in S^a according to P_L.

If S is a topological vector space, it makes sense to say that a linear process is continuous (into $L^0(\Omega,P),d)$). A linear process $L: S \to L^0(\Omega,P)$ will be called <u>canonical</u> if P_L on S^a gives outer measure 1 to S'. Here S' is the set of continuous elements of S^a, i.e. S' is the usual topological dual space of S. Neither canonical nor continuous implies the other in general; however, if S is a metrizable space, a canonical linear process must be continuous. In this report we shall mainly consider Banach spaces S.

Now if S and T are two topological vector spaces, a continuous linear map A from S into T will be called L^0-<u>decomposing</u>, $A \in D(S,T)$, iff for every continuous linear process L on T, $L \circ A$ is canonical on S. If S and T are Hilbert spaces, then $D(S,T)$ is exactly the class of Hilbert-Schmidt operators from S into T (Sazonov [10], Minlos [9]; for a short proof of the harder inclusion see Kolmogorov [8]). The definition of Hilbert-Schmidt operator can be adapted to more general spaces, but the Minlos-Sazonov theorem does not generalize directly. In the next two sections we shall examine some ways of defining sufficiently "small" operators yielding some necessary or sufficient conditions for an operator to belong to $D(S,T)$.

2. <u>Spaces of sequences</u>. As usual, for $1 \leq p < \infty$, ℓ_p denotes the Banach space of all sequences $\{a_n\}_{n=1}^{\infty}$ of real numbers such that

$$||\{a_n\}||_p = (\Sigma|a_n|^p)^{1/p} < \infty .$$

ℓ_∞ is the space of bounded sequences with supremum norm $\|\cdot\|_\infty$ and c_o its closed subspace of sequences converging to 0.

Let B be a set of sequences $\{b_n\}_{n=1}^{\infty}$ of real numbers. Let $1 \leq p \leq \infty$, $1 \leq r \leq \infty$, and suppose that for any $\{b_n\}$ in B, $M_{\{b_n\}}: \{x_n\} \to \{b_n x_n\}$ is a continuous map of ℓ_p into ℓ_r. (If $p \geq r$, this is equivalent to $B \subset \ell_{pr/(p-r)}$.) Then for any topological vector spaces S and T let $(S,\ell_p:B:\ell_r, T)$ denote the set of all compositions $C \circ M_{\{b_n\}} \circ A$ where A is continuous and linear from S into ℓ_p, $\{b_n\} \in B$, and C is continuous and linear from ℓ_r into T. Such an A is defined by an equicontinuous sequence $\{a_n\}$ of elements of S' such that $s \to \|\{a_n(s)\}\|_p$ is a continuous seminorm on S. The latter requirement is redundant for $p = \infty$. A continuous operator C from ℓ_r into T is equivalent to a bounded sequence $\{t_n\}$ in T such that $\{x_n\} \to \Sigma x_n t_n$ is continuous from ℓ_r into T; the latter requirement is redundant for $r = 1$ if T is complete (or quasi- or sequentially complete) and locally convex.

$(S, \ell_\infty : \ell_1 : \ell_1, T)$ is just the set of nuclear operators from S into T as defined by Grothendieck [7]. If $1/p + 1/p' = 1$, then $L^p(S,T)$ in the notation of [5, §9] becomes $(S, \ell_{p'} : \ell_p : \ell_1, T)$ in the current notation. These spaces increase with p. If H and J are Hilbert spaces, the nuclear operators $(H, \ell_\infty : \ell_1 : \ell_1, J)$ are the operators of trace class, and $(H, \ell_2 : \ell_2 : \ell_1, J)$ is exactly the set of Hilbert-Schmidt operators from H to J, which equals $D(H,J)$ as mentioned above.

The general assertion $(S, \ell_p : B : \ell_r, T) \subset D(S,T)$ for all locally convex sequentially complete T is equivalent to its special case $S = \ell_p$, $T = \ell_r$; if it holds we shall say $(\ell_p : B : \ell_r) \subset D$.

Laurent Schwartz asked in 1966 whether $(\ell_\infty : \ell_1 : \ell_1) \subset D$, i.e. whether all nuclear operators are L^0-decomposing. A counter-example turned up, and then Schwartz arrived at the following result. Let $\ell_{p,\log}$ be the set of sequences $\{b_n\}$ such that $\Sigma |b_n|^p (1 + |\log|b_n||) < \infty$. Then $(\ell_\infty : \ell_{1,\log} : \ell_1) \subset D$ and $\ell_{1,\log}$ is the largest class of sequences with this property. In other words, $\ell_{1,\log}$ is the class of all sequences $\{a_n\}$ such that for any sequence $\{X_n\}$ of random variables whose convex combinations are bounded in probability, $\Sigma |a_n X_n|$ converges with probability 1 (Schwartz [12,Cor.2]). A crucial sequence $\{X_n\}$ is formed by independent random variables with the same probability distribution having the Cauchy density $1/\pi(1+x^2)$.

In related results, Schwartz proved that if $1 \leq p < 2$, $\ell_{p,\log}$ is the largest class B of sequences such that $(\ell_{p'}:B:\ell_p) \subset D$. For $p = 2$, as noted above, the log disappears.

If H is an infinite-dimensional Hilbert space, then ℓ_1 is the largest class B of sequences such that $(S,\ell_\infty:B:\ell_2,H) \subset D(S,H)$ for all S ([5,Prop. 8.5], [4,Prop. 6.6]).

3. Gaussian processes (for further details see [4]). A stochastic process $\{x_t, t\varepsilon T\}$ on any set T is called Gaussian iff the joint probability distribution of each set $\{X_t, t\varepsilon F\}$ for F finite, $F \subset T$, is Gaussian. For simplicity, we consider only processes with mean 0, $Ex_t \equiv 0$. Now on a Hilbert space H, there is an "isonormal" Gaussian linear process G which preserves inner products, $EG(x)G(y) = (x,y)$, and this uniquely determines (with $EG(x) \equiv 0$) the joint distributions of $G(x)$ for x in finite sets. Hence the probability measure P_G on H^a or random linear functional is also uniquely determined. With H of infinite dimension, G is continuous but not canonical.

Also, every Gaussian process with mean 0 can be "factored through" an isonormal process, since the natural injection of Gaussian random variables in an $L^0(\Omega,P)$ into $L^2(\Omega,P)$ has an inverse which is isonormal. In this sense there is only one Gaussian process G, which may be restricted to various subsets of H or composed with various functions.

A continuous linear map A from a Banach space S into a Hilbert space H will be called G-decomposing, $A \in GD(S,H)$, iff $G \circ A$ is canonical. Of course $D(S,H) \subset GD(S,H)$; the converse inclusion holds if S is also Hilbert, but the map $\{x_n\} \to \{n^{-1/4}x_n\}$ is in $GD(\ell_1,\ell_2,)$, not in $D(\ell_1,\ell_2)$ ([5, Prop. 8.6]; cf. [4, Prop. 6.7]).

If $(\ell_p,\ell_p:B:\ell_r,H) \subset GD(\ell_p,H)$, then $(S,\ell_p:B:\ell_r,H) \subset GD(S,H)$ for all S and we say $(\ell_p:B:\ell_r,H) \subset GD$. Then $(\ell_\infty:B:\ell_2,H) \subset GD$ iff $B \subset \ell_1$ [4, Prop. 6.6];

$(\ell_2:B:\ell_2,H) \subset GD$ iff $B \subset \ell_2$; and $(\ell_1:B:\ell_2,H) \subset GD$ iff the sequences in B are all $O((\log n)^{-1/2})$ [4, Prop. 6.9].

To fit the above results, especially the latter one, into a consistent pattern leads us away from sequence spaces toward a more precise study of operators. Given a continuous linear map A from a Banach space S into a Hilbert space H, let $C = A(S_1)$ be the image in H of the closed unit ball S_1 in S. Conditions for A to be G-decomposing seem to be connected with the "size" of the convex symmetric closed set C, as measured in suitable ways. (In order that $C \varepsilon GD(S,H)$ it is necessary, although generally far from sufficient, that C be compact, i.e., that A be a compact operator.)

(I) Volumes. Let $V_n(C)$ be the supremum of Lebesque n-dimensional volumes $\lambda_n(P_n C)$ of n-dimensional orthogonal projections P_n of C. Let $EV(C) = \limsup_{n\to\infty} (\log V_n)/n \log n$.

(II) ε-entropy. For $\varepsilon > 0$ let $N(C,\varepsilon)$ be the minimal number of sets of diameter ≤ 2 which cover C. Let

$$r(C) = \limsup_{\varepsilon \downarrow 0} ([\log \log N(C,\varepsilon)]/(\log 1/\varepsilon)).$$

Then $A \notin GD(S,H)$ if $EV(C) > -1$, as is proved in [4, Theorem 5.3] by the following method. Let B be a subset of R^n and let B^0 be its polar,

$$B^0 = \{y \varepsilon R^n : | \langle x,y \rangle | \leq 1 \text{ for all } x \varepsilon B\} .$$

Then $\lambda_n(B)\lambda_n(B^0) \leq v_n^2$ where $v_n = \lambda_n(H_1)$ and H_1 is the ordinary unit ball in R^n, as was proved by L. A. Santalò using integral geometry. Thus if the volumes $\lambda_n(P_n C)$ are too large, then the volumes and hence the Gaussian measures of the polars $(P_n C)^0$ are too small and $A \notin GD(S,H)$.

In the converse direction, $A \varepsilon GD(S,H)$ whenever $EV(C) < -3/2$ and in all known cases where $EV(C) < -1$. Also $r(C) < 2$ implies $A \varepsilon GD(S,H)$ [4, Cor. 3.2].

$A \notin GD(S,H)$ if $r(C) > 2$, according to a result announced by V.N. Sudakov at the 1966 International Congress but only recently published [17].

The relation $r(C) = -2/(1+2EV(C))$ holds in the known cases where $r(C) < 2$. So far it has not been proved to hold generally for r or EV in any interval.

4. Empirical measures. Let (S,d) be a separable metric space. Let P be a Borel probability measure on S. Let $X_1, X_2, \ldots,$ be independent S-valued random variables with distribution P and let P_n be the probability measure with mass $1/n$ at X_j for $j = 1, \ldots, n$. Then with probability 1, P_n converges to P as $n \to \infty$ in the sense that, for every bounded continuous real-valued function f on S, $\int f dP_n \to \int f dP$ [18]. If we know P_n (a finite sample) but do not know P, then estimates of the speed of convergence of P_n to P may help in estimating P.

$\int f dP_n$ converges to $\int f dP$ uniformly for f in any equicontinuous and uniformly bounded set of functions. For instance, let BL_1 be the class of bounded Lipschitzian functions f with

$$\sup \{|f(x)| + |f(x)-f(y)|/d(x,y): x \neq y\} \leq 1;$$

then $E \sup \{|\int f d(P_n - P)|: f \varepsilon BL_1\}$ is of the order of $n^{-1/k}$ where k is the "dimension" of P, if $k > 2$. The dimension k is defined as the infimum of numbers $b > 0$ such that for some $K < \infty$, for all ε with $0 < \varepsilon \leq 1$, S can be covered by at most $K\varepsilon^{-b}$ sets of diameter $\leq 2\varepsilon$ and one set of P-measure at most $\varepsilon^{b/(b-2)}$. (If e.g. P is Lebesque measure on the unit n-dimensional cube, $k = n$.) For the proof see [6, Theorem 3.2].

For any $f_1, \ldots, f_r$ in $L^2(S,P)$, $\{n^{1/2} \int f_j d(P_n - P)\}_{j=1}^r$ converges in law, by the central limit theorem in R^r, to $\{L(f_1), \ldots, L(f_r)\}$ where L is a Gaussian linear process with $EL(f) \equiv 0$ and

$$EL(f)L(g) = \int fg \, dP - \int f dP \int g dP, \quad f, g \varepsilon L^2(S,P) .$$

Let $(T, \|\cdot\|_T)$ be a Banach subspace of the space of all bounded continuous real functions on S with $\sup |f| \leq \|f\|_T$ for all $f \varepsilon T$. Let T' be the dual space of T and let A be the natural map of T into $L^2(S,P)$. We may ask:

a) Is $L \circ A$ canonical, i.e., defined by a countably additive probability measure $P_{L \circ A}$ on T'?

b) If so, does the distribution Q_n of $n^{1/2}(P_n - P)$ in T' converge to $P_{L \circ A}$, i.e., does $\int g \, dQ_n \to \int g \, dP_{L \circ A}$ for every bounded continuous real function g on T'?

a) and b) together would form a central limit theorem in T', implying that if T_1 is the unit ball of T,

$$\sup \{|\int f \, d(P_n - P)| : f \varepsilon T_1\}$$

is roughly of order $n^{-1/2}$. V. Strassen and the author have shown [16] that both a) and b) hold under a suitable ε-entropy hypothesis. For example let S be the unit cube in R^k and for $r > 0$ let T_1 be the set of continuous functions whose partial derivatives of orders $\leq r$ are all bounded by 1. If q is the greatest integer $< r$, the qth derivatives are required to satisfy a Hölder condition of order $r - q$. If $r > k/2$, then the central limit theorem holds, i.e., a) and b) both hold for any P, while if $r \leq k/2$ and P is Lebesque measure, a) fails to hold.

If S is (a subset of) a Euclidean space, one can introduce multidimensional distribution functions (for P_n and P) and study their convergence ([1], [2], [3]).

Work is in progress on finding measures of the distance from P_n to P which are relatively easy to compute while converging to 0 as $n \to \infty$ at a rate also suitable for computation.

REFERENCES

1. Bickel, Peter J., "A distribution free version of the Smirnov two sample test in the p-variate case," Ann. Math. Statist. 40 no. 1 (1969) 1-23.

2. Dudley, R.M., "Weak convergence of probabilities on non-separable metric spaces and empirical measures on Euclidean spaces," Ill. J. Math. 10 (1966) 109-126.

3. ______, "Measures on non-separable metric spaces," Ill. J. Math. 11 (1967) 449-453.

4. ______, "The sizes of compace subsets of Hilbert space and continuity of Gaussian processes," J. Functional Analysis 1 (1967) 290-330.

5. ______, "Random linear functionals," Trans. Amer. Math. Soc. 136 (1969) 1-24.

6. ______, "The speed of mean Glivenko-Cantelli convergence," Ann. Math. Statist. 40 no. 1 (1969) 40-50.

7. Grothendieck, Alexandre, "Produits tensoriels topologiques et espaces nucléaires," Mem. Amer. Math. Soc. no. 16 (1955).

8. Kolmogorov, A. N., "A note to the papers of R. A. Minlos and V. Sazonov," Theor. Probability Appl. 4 (1959) 221-223.

9. Minlos, R.A., "Generalized random processes and their extension to measures," Trudy Moskov. Mat. Obšč. 8 (1959) 497-518; Selected Transl. Math. Stat. and Prob., vol. 3, 1963, 291-314.

10. Sazonov, V., "On characteristic functionals," Theor. Probability Appl. 3 (1958) 188-192.

11. Schwartz, Laurent, "Extension du théorème de Sazonov-Minlos a des cas non hilbertiens," C. R. Acad. Sci. Paris 265 (1967) Ser. A, 832-834.

12. ______, "Réciproque du théorème de Sazonov-Minlos," C. R. Acad. Sci. Paris 266 (1968) Ser. A, 7-9.

13. ______, "Démonstration de deux lemmes sur les probabilités cylindriques," ibid. 50-52.

14. ______, "Un théorème sur les suites de variables aléatoires," Istituto Nazionale di Alta Matematica, Symposia Matematica, Volume II (1968) 203-209.

15. ______, Radon Measures on Arbitrary Topological Spaces, to be published by the Tata Institute of Fundamental Research, Bombay, India.

16. Strassen, Volker, and R. M. Dudley, "The central limit theorem and ε-entropy," Proceedings of an International Symposium on Information Theory and Probability at McMaster Univ., 1968; Springer-Verlag Lecture Note Series.

17. Sudakov, V.N., "Gauss and Cauchy measures and ε-entropy" (in Russian), Doklady Akad. Nauk SSSR 185 (1969) No. 1, pp. 51-53.

18. Varadarajan, V.S., "On the convergence of sample probability distributions," Sankhya 19 (1958) 23-26.

MASSACHUSETTS INSTITUTE OF TECHNOLOGY

Absolute continuity of stochastic processes

Jacob Feldman

A. Introduction

Let $\{X_t, t \in T\}$ be a stochastic process on $(\Omega, \mathcal{F}, P)$ whose range space is a measurable space $(S, \mathcal{S})$. Two stochastic processes X and Y with the same parameter set and the same range space are called stochastically equivalent, $X \approx Y$, if their finite-dimensional distributions are the same: i.e. for each $t_1, \ldots, t_n$ in T and $A_1, \ldots, A_n$ in $\mathcal{S}$, $P\{X_{t_1} \in A_1, \ldots, X_{t_n} \in A_n\} = Q\{Y_{t_1} \in A_1, \ldots, Y_{t_n} \in A_n\}$, Q being the probability measure associated with Y. For each X a certain canonical stochastically equivalent process $\tilde{X}$ exists, whose probability space is $(S^T, \mathcal{S}^T)$ equipped with the measure μ_X which assigns to a set of the form $\{\vec{s} \in S^T : \vec{s}(t_1) \in A_1, \ldots, \vec{s}(t_n) \in A_n\}$ the probability $P\{X_{t_1} \in A_1, \ldots, X_{t_n} \in A_n\}$. The process is then $\tilde{X}_t(\vec{s}) = \vec{s}(t)$. $\tilde{X}$ is called the path space version of X. It is easy to see that $X \approx Y \iff \mu_X = \mu_Y$.

We will say $Y \prec X$ if $\mu_Y \prec \mu_X$ ("absolute continuity"); $Y \sim X$ if both $X \prec Y$ and $Y \prec X$ ("equivalence"); and $X \| Y$ if μ_X and μ_Y have no common absolutely continuous component other than the zero measure ("disjointness"). Disjointness amounts to the existence of disjoint sets Λ_X, Λ_Y in $\mathcal{S}^T$ with $\mu_X(\Lambda_X) = \mu_Y(\Lambda_Y) = 1$.

Suppose $\{f_t, t \in T\}$ is a family of functions: $\Omega \to S$, and $\mathcal{F}$ is the σ-algebra generated by $\{f_t^{-1}(A) : A \in \mathcal{S}\}$. Let P, Q be two probability measures on $\mathcal{F}$ and X, Y the stochastic processes made from $\{f_t, t \in T\}$ by the measures P, Q respectively. Then $Q \prec P \Leftrightarrow \mu_Y \prec \mu_X$, and similarly for the other notions.

A more "algebraic" formulation of absolute continuity may be given: let $\mathcal{M}(X)$ be the algebra of all real-valued random variables which are measurable with respect to the σ-algebra generated by the X_t. Similarly define $\mathcal{M}(Y)$. Then $X \prec Y \Leftrightarrow \exists$ a homomorphism ϕ from $\mathcal{M}(X)$ onto $\mathcal{M}(Y)$, continuous in probability, and sending $1_A(X_t)$ to $1_A(Y_t)$ for each $A \in \mathcal{S}$. Here 1_A is the indicator function of the set A. ϕ is obviously unique. $X \sim Y \Leftrightarrow \phi$ is an <u>isomorphism</u> (in which case continuity is automatic in both directions). It is also possible, but somewhat more complicated, to describe disjointness in these terms. Finally, $X \approx Y \Leftrightarrow \phi$ carries across expectations for bounded members of $\mathcal{M}(X)$.

If $Y \prec X$ then there is a unique nonnegative random variable $J \in \mathcal{M}(X)$ satisfying $E_P\{\xi J\} = E_Q\{\phi(\xi)\}$ for all ξ such that either side exists. J is just the Radon-Nikodym derivative with respect to P of the pullback of Q by ϕ. In particular, if X and Y are defined on the same $(\Omega, \mathcal{F})$, but with possibly different P and Q, then ϕ is induced by the identity map on $\mathcal{F}$-measurable functions, and $J = \frac{dQ}{dP}$.

These notions have a variety of applications:

(1) Hypothesis testing for stochastic processes: if $Y \prec X$, then it is of interest to know the sets $\{J < \text{const}\}$, for maximum likelihood estimates. In case $X \| Y$, the correct hypothesis can of course be chosen with probability 1. A case of importance is when X is a

random noise and Y is a signal plus noise. See Grenander [1] and Root [1] for more details.

(2) Quantum field theory: there is a correspondence between commutation relations for infinitely many variables and equivalence classes of certain stochastic processes; the processes in question have the property that translation by certain constants sends them to equivalent processes. See Gårding and Wightman [1] or Segal [1] for details.

(3) Evaluation of functional integrals: here, as in integration on finite-dimensional spaces, it is important to be able to make absolutely continuous "changes of variable" in order to facilitate evaluation. This sort of "infinite-dimensional calculus" is still in its infancy. The general scheme is to see how transformations on the "observed values" of the process link up with changes of the underlying probability measure. See, for example, Donsker [1].

There are four classes of processes about which we have a sizeable amount of information concerning absolute continuity within the class: sequences of independent random variables, Gaussian processes, decomposable processes, and Markoff processes. I will discuss only the first three. A more detailed survey of all four classes, but with somewhat different emphasis, has been published by Gikhman and Skorokhod [1].

B. Product spaces and sequences of independent random variables

Let $\mathcal{A}_\alpha$ be an increasingly directed set of σ-algebras whose union generates the σ-algebra $\mathcal{A}$. Let μ and ν be probability measures on $\mathcal{A}$, and μ_α, ν_α their restrictions to $\mathcal{A}_\alpha$. An easy version of the Martingale convergence theorem gives us $\frac{d\nu_\alpha}{d\mu_\alpha} \longrightarrow \frac{d\nu}{d\mu}$ in $L_1(\nu)$, provided $\nu \prec \mu$. From this one quickly deduces:

(4) in general, $\sqrt{\frac{d\nu_\alpha}{d\mu_\alpha}} \longrightarrow \sqrt{\frac{d\nu}{d\mu}}$ weakly in $L_2(\nu)$.

Corollaries:

(5) $\nu \prec \mu \iff \left(\sqrt{\frac{d\nu_\alpha}{d\mu_\alpha}}, \sqrt{\frac{d\nu_\beta}{d\mu_\beta}}\right) \longrightarrow 1$ (inner product in $L_2(\nu)$),

(6) $\nu \| \mu \iff \left(\sqrt{\frac{d\nu_\alpha}{d\mu_\alpha}}, 1\right) \longrightarrow 0$.

Now let our space be S^T, $\mathcal{A} = \mathcal{S}^T$, let our directed index set be the finite subsets $F \subset T$, and $\mathcal{A}_F$ the σ-field generated by the coordinates in F. Let μ_t, ν_t be probability measures on $\mathcal{S}$ for each $t \in T$, and we will abuse the language by also letting them be the corresponding measures on $\mathcal{S}_{\{t\}}$ when convenient. Let $\mu = \prod_t \mu_t$, $\nu = \prod_t \nu_t$. Then μ_F is the measure μ_F whose projection onto $(S^F, \mathcal{S}^F)$ is $\prod_{t\in F} \mu_t$, and similarly for ν_F. So $\frac{d\nu_F}{d\mu_F} = \prod_{t\in F} \frac{d\nu_t}{d\mu_t}$ (this is quite general, even without any absolute continuity, provided $\frac{d\nu_t}{d\mu_t}$ is defined as the derivative with respect to μ_t of that part of ν_t which is $\prec \mu_t$). Then $\left(\sqrt{\frac{d\nu_F}{d\mu_F}}, 1\right) = \prod_{t\in F} \int \sqrt{\frac{d\nu_t}{d\mu_t}}\, d\mu_t$, and $\mu \| \nu \iff \prod_{t\in F} \int \sqrt{\frac{d\nu_t}{d\mu_t}}\, d\mu_t = 0$. Now we prove a slight rephrasal of the main theorem of Kakutani [1]:

(7) If $\mu \not\parallel \nu$, then $\nu \prec \mu \Leftrightarrow$ each $\nu_t \prec \mu_t$. Indeed,

$$\left(\sqrt{\frac{d\nu_F}{d\mu_F}}, \sqrt{\frac{d\nu_G}{d\mu_G}}\right) = \left(\prod_{t\varepsilon F\cap G} \int \frac{d\nu_t}{d\mu_t}\, d\mu_t\right)\left(\prod_{t\varepsilon F\Delta G} \int \frac{d\nu_t}{d\mu_t}\, d\mu_t\right),$$

where Δ means symmetric difference. If some ν_{t_o} is not $\prec \mu_{t_o}$,

then the factor $\int \frac{d\nu_{t_o}}{d\mu_{t_o}}\, d\mu_{t_o} < 1$, and all other factors are ≤ 1 ,

so since t_o is eventually in $F \cap G$, the product cannot $\to 1$.

On the other hand, if each $\nu_t \prec \mu_t$, then the first factor is 1 ,

while the fact that $\prod_{t\varepsilon F} \int \sqrt{\frac{d\nu_t}{d\mu_t}}\, d\mu_t \not\to 0$ implies that the second

factor $\to 1$. It is also worth noting that $\frac{d\nu}{d\mu} = \prod_{t\varepsilon T} \frac{d\nu_t}{d\mu_t}$, conver-

gence occurring for example weakly in $L_1(\nu)$.

Let $\{X_t, t \varepsilon T\}$ be the stochastic process obtained from the coordinate functions on $(S^T, \mathcal{S}^T, \mu)$ and $\{Y_t, t \varepsilon T\}$ that obtained by using ν . Here is an elegant result from Shepp [1] that has been obtained in this context. Suppose all μ_t are the same μ_o , that S is the real line, and that $\mu_o(ds) = g(s)ds$, where $g > o$ a.e. Let $\nu_t(A) = \mu_o(A-m_t)$, so $\nu_t(ds) = g(s-m_t)ds$. This last amounts to saying $Y \approx \{X_t+m_t, t \varepsilon T\}$. So each $\nu_t \sim \mu_t$. Then it is shown using Kakutani's theorem that $\sum_t |m_t|^2 = \infty \Rightarrow X \| Y$. As for the converse, that $\sum_t |m_t|^2 < \infty \Rightarrow X \sim Y$, this will hold $\Leftrightarrow \int |x \, \widehat{\sqrt{g}}\,(x)|^2 dx < \infty$, where $\widehat{\sqrt{g}}$ is the Fourier transform of $\sqrt{g}$. See also Dudley [1] for a generalization of this. In general, the problem of absolute continuity of $\{X_t + m_t , t \in T\}$ with respect to X is an important one, as is clear from the interpretation of m as a signal and X as a random noise.

C. Gaussian processes

A real-valued random variable ξ will be called Gaussian if its distribution has the density $\frac{1}{\sqrt{2\pi c}} e^{-\frac{(x-m)^2}{2c}}$ for some $c > 0$ and some m ; or also if ξ is with probability 1 a constant m. In either case the mean $E\{\xi\} = m$, while the variance $E\{(\xi-m)^2\} = c$ in the first case and 0 in the second. A random variable ξ with values in a locally convex vector space S will be called Gaussian if $s^*(\xi)$ is Gaussian for each $s^* \in S^*$. A stochastic process $\{X_t, t \in T\}$ will be called Gaussian if each finite linear combination $\sum_t a_t X_t$ is Gaussian. There is no loss of generality in the present section if we stick to real-valued processes.

A real-valued Gaussian process is determined up to stochastic equivalence by the function $t \to m_t = E\{X_t\}$, its mean, and the function $t, u \to C(t,u) = E\{(X_t - m_t)(X_u - m_u)\}$, its covariance. C is nonnegative definite. Furthermore, any real-valued function m and nonnegative definite function C are the mean and covariance of a real-valued Gaussian process. See Doob [1].

Let X be a Gaussian process with mean 0, and let $\mathcal{H}_X$ be the set of all limits in probability of finite linear combinations of the X_t. Then $\mathcal{H}_X$ consists entirely of Gaussian random variables with mean 0, and is a closed subspace of $L_2(P)$. Consider the problem of composing X with $X+m = \{X_t + m_t, t \in T\}$. This was essentially solved by Grenander [1], and the solution can be obtained immediately from Kakutani's theorem:

(8) $X \sim X + m \Longleftrightarrow$ there is an element $\xi \in \mathcal{H}_X$ such that $m_t = (\xi, X_t)$ (the inner product is that of $\mathcal{H}_X \subset L_2(P)$). Otherwise, $X \| X + m$.

If $X \sim X + m$, then $J = e^{\xi - \frac{\|\xi\|^2}{2}}$.

Now let Y be another Gaussian process with mean 0. Then a variety of proofs exist that either $X \sim Y$ or $X \| Y$; see Feldman [1], Hajeck [1], Shepp [2] for example. Necessary and sufficient for $X \sim Y$, as given in Feldman [1] (see also Segal [1]) are the following. An <u>equivalence operator</u> H from Hilbert space $\mathcal{H}$ to Hilbert space $\mathcal{K}$ is a linear homeomorphism from $\mathcal{H}$ onto $\mathcal{K}$ such that $\sqrt{H^*H} - I$ is Hilbert-Schmidt. Then:

(9) $X \sim Y \Longleftrightarrow \exists$ an equivalence operator $H : \mathcal{H}_X \to \mathcal{H}_Y$ such that $H(X_t) = Y_t$ for all $t \in T$.

In this case, J may be computed as follows. Let $\{\xi_i : i \in I\}$ be a c.o.n.s. in $\mathcal{H}_X$ such that $\{H\xi_i : i \in I\}$ are orthogonal. Let $\eta_i = H\xi_i$ and $c_i = \|\eta_i\|^2$. Let $\eta = \{\eta_i, i \in I\}$, $\xi = \{\xi_i, i \in I\}$, regarded as stochastic processes. Then μ_ξ is a product measure on S^I, its ith factor having density $\frac{1}{\sqrt{2\pi}} e^{-\frac{|s|^2}{2}}$, while μ_η is a product measure whose ith factor has density $\frac{1}{\sqrt{2\pi c}} e^{-\frac{|s|^2}{2c}}$.

From this it is easy to see that $J = \prod_i \frac{1}{\sqrt{e_i}} e^{-\frac{1}{2}(\frac{1}{e_i} - 1)\xi_i^2}$. Notice that in the case of the shift of the mean, log J was a first degree expression in X , i.e. a constant plus an element of $\mathcal{H}_X$, while in the present case log J is quadratic in X .

Finally, we compare X + m with Y + n . Shifting both processes by the same mean leaves absolute continuity relations unaffected, so it may be assumed that n = 0 . The result is that $X + m \perp Y \iff X \sim Y$ and $X \sim X + m$. This may now be seen by considering the path space $(S^T, \mathcal{S}^T)$, and the ascending σ-subalgebras $\mathcal{a}_F$ of $\mathcal{S}^T$:

$$\left(\sqrt{\frac{d\mu_{Y,F}}{d\mu_{X,F}}}, 1\right) \geqq \left(\sqrt{\frac{d\mu_{Y-m,F}}{d\mu_{X,F}}}, 1\right) ,$$

inner products being taken in $L_2(\mu_X)$. This is just a calculation with multidimensional Gaussian densities. So if the terms on the left → o so do those on the right, and (5) and (6) may be applied.

The condition (9) above may be rephrased in terms of the covariances C and D of X and Y . Here are some cases where these conditions turn out to be manageable.

(10) T is a locally compact abelian group, C(t,u) = c(t-u) , D(t,u) = d(t-u) , where c and d are continuous positive-definite functions on T and hence the Fourier transforms of measures γ and δ on the dual group. Necessary and sufficient for X ~ Y is that γ and δ have identical continuous parts, and their points of mass occur at the same points, while the corresponding masses γ_i and δ_i satisfy $\sum_i \left(\frac{\delta_i}{\gamma_i} - 1\right)^2 < \infty$. See Feldman [1].

(11) T is the interval $[-\tau,\tau]$, $C(t,u) = c(t-u)$, $D(t,u) = d(t-u)$ for $|t|$, $|u| \leqq \tau$, where c and d are again continuous positive-definite functions, Fourier transforms of measures γ and δ on the real line. Let γ have the special form $\gamma(dx) = g(x)dx$, where g is bounded above and below by constant nonzero multiples of $\frac{1}{(1+x^2)^n}$, n a fixed positive integer. Then necessary and sufficient conditions on Y that $Y \sim X$ are (i) the function $e = c - d$ has $2n - 1$ derivatives, the $(2n-1)$st being absolutely continuous, so that $\left(1 - \frac{d^2}{dt^2}\right)^n e$ exists a.e., and

$$\int_{-\tau}^{\tau}\int_{-\tau}^{\tau} \left|\left(1 - \frac{d^2}{dt^2}\right)^n e\right|^2 < \infty .$$

(ii) if f is an entire function of exponential type $\leqq \tau$, and $0 < \int |f(x)|^2 \frac{1}{(1+x^2)^n} dx < \infty$, then also $\int |f(x)|^2 \delta(dx) > 0$. See Feldman [2], [3], Rosanov [1]. It would be of great interest if conditions could be obtained without (*).

(12) T is $[0,\tau]$, where $0 < \tau \leqq \infty$, and X is the Wiener process starting at 0, so that $C(t,u) = \min(t,u)$. Then necessary and sufficient for $X \sim Y$ are (i) $E = C - D$ is the indefinite integral in $T \times T$ of a Lebesque square-integrable function, (ii) the integral operator on $L_2(T)$ obtained from E does not have 1 in its spectrum. See Shepp [3], Golosov [1]. In the latter paper, in fact, conditions are obtained when X is an arbitrary Markov Gaussian process.

Finally, let us briefly consider the effect of transforming the process. Translations have already been considered. Now let each Y_t

be a finite linear combination of the X_t. Thus, $Y_t = H_o X_t$, H_o being a linear transformation defined on the space of finite linear combinations of the X_t. Then it turns out that necessary and sufficient for $Y \sim X$ is that H_o be the restriction of an equivalence operator $H : \mathcal{H}_X \rightarrow \mathcal{H}_X$. Furthermore, every zero mean Gaussian $Y \sim X$ is stochastically equivalent to one obtained in this manner. See Segal [1]. Nonlinear transformations have also been considered; there are some quite difficult although still not completely satisfying results in Gross [1].

D. Decomposable processes and continuous products of probability spaces

Natural generalizations of sequences of independent vector-valued random variables are the vector-valued decomposable processes. We shall restrict ourselves to the continuous case. For present purposes, S will be a separable Hilbert space, $\mathcal{S}$ its Borel sets. T will be the unit interval, and our processes will be parameterized by the elements of $\mathcal{B}$, the Borel sets of T. An S-valued process X will be called <u>continuously decomposable</u> if

(i) $B_1,\ldots,B_n$ disjoint $\Rightarrow X_{B_1},\ldots,X_{B_n}$ independent,

(ii) $B_1,B_2,\ldots$ disjoint with union $B \Rightarrow \sum_i X_{B_i} = X_B$ with probability one,

(iii) $X_{\{t\}} = 0$ with probability one for each $t \in T$.

It is known (Shale-Stinespring [1], Feldman [4]) that such an X is infinitely divisible; in fact its characteristic functional has a logarithm of the form

(13) $\log E\{e^{i(X_B,r)}\} = i(\alpha(B)r,r) - \frac{(C(B)v,v)}{2} + \int_S k(r,q)\Pi\ (B\times dq)$,

where α is an S-valued measure on $\mathcal{B}$, C is a measure on $\mathcal{B}$ whose values are nonnegative trace-class operators on S , Π is a nonnegative numerical measure on $\mathcal{B} \times \mathcal{S}$ satisfying certain requirements, and k is the kernel $k(r,q) = e^{i(r,q)} - 1 - \frac{i(r,q)}{1+\|q\|^2}$.

If Y is another continuously decomposable process with the same $(S,\mathcal{S})$ and $(T,\mathcal{B})$, when is $Y \prec X$? We shall see what J must be like. The solution is mainly due to Skorokhod, as described in Skorokhod-Gikhman [1].

Let $\mathcal{F}_{B_o}(X)$ be the σ-field generated by $\{X_B : B \subset B_o\}$, where $B_o \in \mathcal{B}$. Then by restricting X and Y to subsets of B_o , one gets a function J_{B_o} , which may also be interpreted as $E\{J|\mathcal{F}_{B_o}\}$. J itself is of course J_T . It is easy to see that if $B_1,B_2,\ldots$ are disjoint with union B then $J_B = \prod_{i=1}^{\infty} J_{B_i}$. Thus it is a matter of finding all such "miltiplicatively decomposable" nonnegative real-valued processes with each J_B being $\mathcal{F}_B(X)$ -measurable and $E\{J_B\} = 1$ for all $B \in \mathcal{B}$.

The family $\{\mathcal{F}_B(X)\ ,\ B \in \mathcal{B}\}$ of σ-fields on the probability space (Ω, P) of X satisfies the following conditions:

(i) $B_1,\ldots,B_n$ disjoint $\Rightarrow \mathcal{F}_{B_1}(X),\ldots,\mathcal{F}_{B_n}(X)$ independent,

(ii) $B_1,B_2,\ldots$ disjoint with union $B \Rightarrow \bigvee_{i=1}^{\infty} \mathcal{F}_{B_i}(X) = \mathcal{F}_B(X)$ (up to sets of probability zero),

(iii) $\mathcal{F}_{\{t\}}(X)$ contains only sets of probability zero and one, for each $t \in T$.

A system $\Phi = \{\mathcal{F}_B : B \in \mathcal{B}\}$ of σ-fields of measurable sets in a probability space satisfying (i), (ii), (iii) will be called a continuously factored probability space. This provides a continuous analogue of a product of probability spaces. A continuously decomposable process Y will be called Φ-measurable if Y_B is $\mathcal{F}_B$-measurable for all $B \in \mathcal{B}$. The system $\{\mathcal{F}_B(X) : B \in \mathcal{B}\}$ of the previous paragraph will be called $\Phi(X)$; thus, X is $\Phi(X)$-measurable.

Question. Is every Φ a $\Phi(X)$ for some X ? Or even: does every Φ carry a nontrivial Φ-measurable decomposable process? The answer is unknown. However, there is a tie-up with absolute continuity. Say a measure $Q \prec P$ is admissable if the $\mathcal{F}_B$ remain independent for disjoint B with respect to the measure Q as well. If $\Phi = \Phi(X)$, then this is precisely the condition that X remain a continuously decomposable process under Q .

Fact. Φ possesses some admissable $Q \prec P \iff \Phi$ possesses some admissable $Q \prec P \iff \Phi$ carries some nontrivial decomposable process.

The proof essentially rests on the fact that if $Q \sim P$ then

$B \longrightarrow \log \dfrac{d(Q|\mathcal{F}_B)}{d(P|\mathcal{F}_B)}$ is a decomposable process.

We shall be interested in the case when there are enough Φ-measurable decomposable Y to generate Φ , i.e. each $\mathcal{F}_B$ is generated by $\{Y_B : Y$ a Φ-measurable decomposable process$\}$. Such a Φ will be called linearizable. It is shown in Feldman [1], [2] that any linearizable Φ is a $\Phi(X)$ for some X .

Any continuous decomposable process X may be written as $X' + X''$, where X' and X'' are $\Phi(X)$-measurable decomposable processes, independent of each other, X' being Gaussian with mean zero and X'' a mixture of generalized Poisson processes (about which more later; see Loeve [1], Feldman [1], [2]); this is part of the Levy-Ito theorem. Correspondingly, any linearizeable Φ factors into $\Phi' \times \Phi''$ (the product being defined in a natural fashion), where Φ' carries only Gaussian decomposable processes while Φ'' carries only mixed Poissonian decomposable processes. If $\Phi = \Phi(X)$, where $X = X' + X''$ as above, then the Φ' in question is $\Phi(X')$ and Φ'' is $\Phi(X'')$.

For the Gaussian case, $\Phi(X')$, the only admissable measures arise from a shift of the mean of X'. This may be rephrased by saying: if X' is a Gaussian continuously decomposable process, and $Y \prec X'$, then $Y \sim X' + m$ for some measure m on $\mathcal{B}$. This is easy to see from the results of part C, and furthermore the exact form of m and of J can be deduced from that section.

Thus, the general problem is reduced to the Poissonian case, and hereafter we assume that Φ is purely Poissonian. Such a Φ may be written as $\Phi(X)$ where X is a mixed Poisson process, i.e. the term $C(B)$ does not occur in formula (13) for $\log E\{e^{i(X_B,r)}\}$; and α may also be assumed zero. Π is called the Levy-Khintchine measure of X. The rest of the Levy-Ito theorem now says

(14) $X_B = \int_B s\{\nu(B \times ds) - \frac{1}{1+\|s\|^2}\Pi(B \times ds)$, where the stochastic integral is defined as in Skorokhod-Gikhman [1]. ν is a generalized Poisson process on $B \times \mathcal{S}$: i.e. if C is a set with $\Pi(C) < \infty$, then $\nu(C)$ is defined, and is a Poisson random variable (with mean $\Pi(C)$), and further, ν satisfies the analogues of (i), (ii).

Now let Q be a $\Phi(X)$-admissable measure $\prec P$. Then there exists a measurable function $f : T \times S \longrightarrow (-\infty) \cup$ (reals), uniquely determined Π - a.e., such that $\int \frac{|e^f - 1|}{1+|e^f-1|^2} d\Pi < \infty$, and

$\log \frac{dQ}{dP} = \int\{fd\nu - (e^f-1)\}d\Pi$. So $Q \sim P \Longleftrightarrow \Pi(\{f = -\infty\}) = 0$. With respect to this new measure, $\log E_Q\{e^{i(X_B,r)}\} = \int_S k(r,q)\, \textcircled{H}\, (B \times dq)$, where $\textcircled{H} \prec \Pi$ and $\frac{d\textcircled{H}}{d\Pi} = e^f$. In particular, notice that if α is an S-valued measure such that $X + \alpha \quad X$, then $\alpha \equiv 0$.

From all this, it is apparent that mixed Poissonian processes are only moderately amenable to absolutely continuous transformations. A natural sort of transformation on X is a stochastic integral: $Y_B = \int_B F(t)X_{dt}$, where $F(t)$ is a linear operator: $S \longrightarrow S$. But, aside from conditions on F to allow the integral to be defined, the conditions for $Y \prec X$ are, setting $\dot{F}(s,t) = (s,F(t))$, that

$\Pi \circ \dot{F}^{-1} \prec \Pi$, and $\log \frac{d(\Pi \circ \dot{F}^{-1})}{d\Pi}$ be an f of the kind described in the theorem. This could be worked out explicitly for the stable processes, for example.

<u>Question</u>. Observe that if X and Y are two mixed Poissonian processes with Levy-Khintchine measures Π and $\textcircled{H}$ respectively, and $\Pi \| \textcircled{H}$, then $\exists$ Z mixed Poissonian, nontrivial, such that $Z \prec X$ and Y both, and therefore $X \| Y$. But suppose on the other hand that $\Pi \| \textcircled{H}$. Does this imply $X \| Z$?

References

M. Donsker

1. On function space integrals, in "Analysis in Function Space," edited by W. T. Martin and I. E. Segal, M.I.T. Press, Cambridge, Mass. 1964.

J. L. Doob

1. "Stochastic Processes," Wiley, N.Y., 1953.

R. F. Dudley

1. Singularity of measures on linear spaces. Z Wahr. und Verw. Gebiete 6 (1966), 129-132.

J. Feldman

1. Equivalence and perpendicularity of Gaussian processes, Pac. J. of Math. 8 (1958), 699-708.

2. A clarification concerning certain equivalence classes of Gaussian processes on an interval. Ann. Math. Stat. 39 (1968), 1078-79.

3. Decomposable products and continuous products of probability spaces, to appear, Proceedings of Conference on Probability Theory on Algebraic Structures, Clermont-Ferrand, 1969.

4. Decomposable processes and continuous products of probability spaces, to appear.

L. Gårding and A. S. Wightman

1. Representations of the commutation relations, Proc. Nat. Acad. Sci. U.S.A. 40 (1954), 622.

I. I. Gikhman and A. V. Skorokhod

1. On the densities of probability measures in function spaces, Russian Math. Surveys, 21 (1966), 83-156.

Yu I. Golosov

1. Gaussian measures equivalent to Gaussian Markov measures. Soviet Math. 7 (1966), 48-52.

U. Grenander

1. Stochastic processes and statistical inference, Arkiv für Matematik, Band 1, mr. 17.

L. Gross

1. Integration and nonlinear transformations in Hilbert space. Trans. Amer. Math. Soc. 94 (1960), 404-440.

J. Hajek

1. On a property of normal distributions of any stochastic process. Czech. Math. J. 8 (1958), 610-618.

M. Loeve

1. "Probability Theory," Van Nostrand, Princeton, N.J., 1960.

Yu. A. Rozanov

1. On the densities of Gaussian measures and Wiener-Hopf's integral equations, Theory of Prob. and its Applications 1 (1966), 161-79.

W. L. Root

1. Singular Gaussian measures in detection theory, Proc. Symp. Time Series Analysis, Wiley, N.Y. (1963).

I. E. Segal

1. Distributions in Hilbert space and canonical systems of operators, Trans. Amer. Math. Soc. 88 (1958), 12-41.

D. Shale and W. F. Stinespring

1. Continuously splittable distributions in Hilbert space, Ill. Jour. Math., 10 (1966), 574-78.

L. A. Shepp

1. Distinguishing a sequence of random variables from a translate of itself, Ann. Math. Stat. 36 (1965), 1107-1112.

2. Gaussian measures in function space, Pacific J. Math., 17 (1966), 167-173.

3. Radon-Nikodym derivatives of Gaussian measures, Ann. Math. Stat., 37 (1966), 321-354.

Quantization and Unitary Representations

by Bertram Kostant

Part I: Prequantization

0. Introduction. 1. This paper is the first part of a two part paper dealing with the question of setting up a unified theory of unitary representations of connected Lie groups.

We have found that when the notion of what the physicists mean by quantizing a function is suitably generalized and made rigorous, one may develop a theory which goes a long way towards constructing all the irreducible unitary representations of a connected Lie group. In the compact case it encompasses the Borel-Weil theorem. Generalizing Kirillov's result on nilpotent groups, L. Auslander and I have shown that it yields all the irreducible unitary representations of a solvable group of type I. (Also a criterion for being of type I is simply expressed in terms of the theory.) For the semi-simple case, by results of Harish-Chandra and Schmid, it appears that enough representations are constructed this way to decompose the regular representation.

The theory is founded in differential geometry. A principal point is that the 2-form of a symplectic manifold under a certain condition (integrality condition) is the curvature of a line bundle with connection; that the Hilbert space involved (as far as this paper is concerned) is to be found among the sections of this line bundle and

---

Work on this paper was partially supported by a grant from the National Science Foundation.

that operating on these sections is a Lie algebra (under Poisson bracket) of functions on the manifold. This assignment mapping functions to operators is quantization.

The extraction of the Hilbert space and the Lie algebra above requires the notion of polarizing the symplectic manifold. (In the classical quantum mechanical situation, this means e.g. isolating the q's from the even dimensional space of p's and q's. The notion, however, is broad enough so that even in this case it also yields the Bargmann-Segal-Fock representation of the Heisenberg Lie algebra on the $z = q + ip$ as well as the usual one on the q's.) We will deal with this in Part II. In Part I we consider only pre-quantization (see § 4.3).

A representation of a group will arise from a symplectic homogeneous space X when the corresponding Lie algebra of hamiltonian vector fields can be lifted to functions which are quantizable. One of the results (Theorem 5.4.1) is that this is the case when and only when X corresponds to or covers an orbit in the dual of the Lie algebra, justifying the idea of finding all the irreducible representations from these orbits (as Kirillov did in the nilpotent case). This fact also yields a generalization of Wang's theorem characterizing compact Kahler homogeneous spaces. Also (Corollary 1 to Theorem 5.7.1) generalizing results of Borel-Weil in the compact case, one has that the 2-form on the orbit, defined by a linear functional on the Lie algebra f, is integral if and only if f is the differential of a character on the isotropy group (which may be disconnected) at f.

Part I is devoted to the differential geometry foundations of the theory. For completeness we include with proofs basic facts (many of which are known, although some appear to be new, e.g. Theorem 2.5.1, classifying all line bundles with connection having a given curvature) in the theory of line bundles with connection. An important point is Theorem 1.13.1 which associates a central group extension to every line bundle with connection. This extension in the symplectic case will later be related to, infinitesmally, the central extension of the hamiltonian vector fields defined by the Lie algebra of functions under Poisson bracket. This relation plays a central role in the idea of pre-quantization.

1.1 All manifolds considered here are assumed to be differentiable of class C^∞, separable (as topological spaces), and unless otherwise indicated, connected. Also the word smooth when applied to maps between manifolds means differentiable of class C^∞. If M is a manifold, $C(M)$ denotes the algebra of all complex-valued smooth functions on M.

If $\pi: M \longrightarrow N$ is a surjective map of sets then a section of π or, just section, if π is understood is a map $s: N \to M$ such that $\pi \circ s$ is the identity map on N.

Let X be a manifold. By a line bundle L over X we mean a vector bundle (in the smooth sense) over X with $\underline{\underline{C}}$, the complex numbers, as fiber. This is denoted schematically by

$$\begin{array}{ccc} \underline{\underline{C}} & \longrightarrow & L \\ & & \Big\downarrow \pi \\ & & X \end{array}$$

Thus L is a manifold. The projection map π is smooth, and if for any $p \in X$ one puts $L_p = \pi^{-1}(p)$, then L_p is a one-dimensional vector space over $\underline{\underline{C}}$. Moreover there exists a (open) covering $\mathcal{U} = \{U_i\}$, $i \in I$, of X and nowhere vanishing smooth sections $s_i \colon U_i \longrightarrow L$ on U_i, $i \in I$, such that the map $\eta_i \colon C \times U_i \longrightarrow \pi^{-1}(U_i)$, given by $\eta_i(\,(z,q)\,) = z\, s_i(q)$, is a diffeomorphism.

A set of pairs $\{(U_i,s_i)\}$, $i \in I$, satisfying the above relations is called a local system for the line bundle over X. Given such a local system, the corresponding set of transition functions are the elements $c_{ij} \in C(U_i \cap U_j)$, $i,j \in I$ defined by $c_{ij}\, s_i = s_j$ in $U_i \cap U_j$. One of course has

$$c_{ij} = c_{ji}^{-1} \quad \text{and} \quad c_{ij}\, c_{jk} = c_{jk} \quad \text{in} \quad U_i \cap U_j \cap U_k\,. \qquad (1.1.1)$$

We denote by $S(X,L)$ or more briefly S if X and L are understood, the space of all smooth sections $s \colon X \longrightarrow L$. If $s \colon X \longrightarrow L$ is a section and ϕ is a complex-valued function on X, we let ϕs be the section given by $(\phi s)(p) = \phi(p)\, s(p)$. The space S is then clearly a $C = C(X)$ module.

Given the local system above and any section (not necessarily continuous) $s \colon X \longrightarrow L$, one uniquely writes $s = \phi_i\, s_i$ on U_i where ϕ_i is a complex-valued function on U_i. The ϕ_i are called the local coordinates of s and of course one has

$$c_{ij}\, \phi_j = \phi_i \qquad \text{on} \quad U_i \cap U_j\,. \qquad (1.1.2)$$

Conversely, any family of functions ϕ_i (on U_i) satisfying (1.1.2) is the set of local coordinates for a unique section s. Clearly $s \in S$ if and only if $\phi_i \in C(U_i)$ for all i. One may consider however the larger space $S_m = S_m(X,L)$ of all measurable sections. A section

$s: X \longrightarrow L$ is called measurable if for all i, ϕ_i is a (Borel) measurable function on U_i. This definition is clearly independent of the choice of the local systems.

1.2. We recall that two line bundles L^1 and L^2 over X are called equivalent if there exists a diffeomorphism $\tau: L^1 \longrightarrow L^2$ such that for any $p \in X$, τ induces a linear isomorphism $L_p^1 \longrightarrow L_p^2$. This defines an equivalence relation in the set of line bundles over X and we let $\mathcal{L} = \mathcal{L}(X)$ be the set of equivalence classes.

One knows that $\mathcal{L}$ has a group structure (using tensor products of line bundles) and that this group is naturally isomorphic with $H^2(X, \underline{\underline{Z}})$. (The theory of the first chern class.) We recall the definition (Cech) of $H^i(X,A)$ for an abelian group A and how the isomorphism is constructed. A covering $\mathcal{U} = \{U_i\}$, $i \in I$, of X is called contractible if $U_{i_0 \ldots i_k} = U_{i_0} \cap \ldots \cap U_{i_k}$ is (smoothly) contractible for every k and every $(i_0, \ldots, i_k) \in I^{k+1}$ such that $U_{i_0, \ldots, i_k} \neq \emptyset$ (i.e. for every simplex). One knows that contractible coverings exist and that in fact every covering has a contractible refinement. (One uses for example, convex neighborhoods relative to any Riemannian structure on X).

Let A be any abelian group and let $\mathcal{U} = \{U_i\}$, $i \in I$, be a covering of X. Relative to the covering a k-chain is an arbitrary function $(i_0, \ldots, i_k) \longrightarrow a_{i_0 \ldots i_k} \in A$ from the set of all k simplices into A. The set of all k-chains from a group $C^k(\mathcal{U}, A)$, using addition of functions, and one obtains a homomorphism $d: C^k(\mathcal{U}, A) \rightarrow C^{k+1}(\mathcal{U}, A)$ by defining, for $a \in C^k(\mathcal{U}, A)$

$$(d\,a)_{i_0\ldots i_{k+1}} = \sum_{j=0}^{k+1} (-1)^j \; a_{i_0\ldots \hat{i}_j \ldots i_{k+1}} \, .$$

One has $d^2 = 0$ and the corresponding cohomology group is denoted by $H^k(\mathcal{U},A)$. If $\mathcal{V}$ is a refinement of $\mathcal{U}$ one has a homomorphism $H^k(\mathcal{U},A) \longrightarrow H^k(\mathcal{V},A)$ enabling one to take, over all coverings, the inductive limit. This limit is the (Cech) cohomology group $H^k(X,A)$ and for any particular covering $\mathcal{U}$ one has a natural homomorphism $H^k(\mathcal{U},A) \longrightarrow H^k(X,A)$. However if $\mathcal{U}$ is contractible then this map is an isomorphism (see Godement [1], Corollary, p. 213, and also cf. Theorem 5.9.2, p. 227). Thus we can identify $H^k(\mathcal{U},A)$ with $H^k(X,A)$ when $\mathcal{U}$ is contractible.

For convenience we note that if $a \in C^2(\mathcal{U},A)$ then it is cocycle if and only if it satisfies the relations

$$a_{kjl} - a_{ikl} + a_{ijl} - a_{ijk} = 0 \qquad (1.2.1)$$

whenever $U_{ijkl} \neq \emptyset$. It then defines an element $[a] \in H^2(X,A)$ which vanishes if and only if there exists $b \in C^2(\mathcal{U},A)$ such that

$$a_{ijk} = b_{ij} + b_{jk} - b_{ik} \, .$$

Now let L be a line bundle over X and let $\{(U_i, s_i)\}$, $i \in I$, be a local system for L where $\{U_i\}$, $i \in I$, (by taking a refinement if necessary) is a contractible covering of X. Since $U_i \cap U_j$, if not empty, is contractible and hence simply connected we can define a smooth function $f_{ij} : U_i \cap U_j \longrightarrow \underline{C}$ by

$$f_{ij} = \frac{1}{2\pi i} \log c_{ij}$$

where the c_{ij} are the transition functions. If $U_i \cap U_j \cap U_k \neq \emptyset$

then since $c_{ij} c_{jk} = c_{ik}$ one clearly has $\exp 2\pi i \ a_{ijk} = 1$ where

$$a_{ijk} = f_{ij} + f_{jk} - f_{ik} .$$

But then a_{ijk} must be $\underline{\underline{Z}}$ - valued and since continuous, it must be constant on U_{ijk} defining an element of $C^2(\mathcal{U},\underline{\underline{Z}})$. But it is clearly a cocycle and therefore defines an element $[a] \in H^2(X,\underline{\underline{Z}})$ which is clearly independent of the choice of logarithms.

That the class $[a] \in H^2(X,\underline{\underline{Z}})$ is also independent of the choice of local system and in fact depends only on the class $[L] \in \mathcal{L}$ of L follows from the fact that if c_{ij}^1 and c_{ij}^2 are the transition functions for the line bundles L^1 and L^2 relative to respective local systems (U_i, s_i^1) and (U_i, s_i^2) , $i \in I$, then L^1 and L^2 are equivalent if and only if there exists smooth functions $\lambda_i: U_i \longrightarrow C^*$ such that

$$\lambda_i \, c_{ij}^1 \lambda_j^{-1} = c_{ij}^2 \qquad \text{on } U_i \cap U_j \tag{1.2.2}$$

One then has a map

$$\kappa: \mathcal{L} \longrightarrow H^2(X,\underline{\underline{Z}}) \tag{1.2.3}$$

where $\kappa[L] = [a]$. The injectivity and surjectivity of κ follows using a partition of unity. That is, choose $\mathcal{U}$ to be both contractible and locally finite. One thus has a partition of unity $\sum_{i \in I} h_i = 1$ where the support of h_i lies in U_i . But now if $a \in C^2(\mathcal{U},\underline{\underline{Z}})$ is a cocycle then an element $f_{ij} \in C(U_i \cap U_j)$ is well defined by

$$f_{ij} = \sum_{p \in I} a_{ijp} \, h_p$$

and one has, by (1.2.1),

$$f_{ij} + f_{jk} - f_{ik} = a_{ijk} \in \underline{\underline{Z}} \quad .$$

If now one puts $c_{ij} = \exp 2\pi i f_{ij}$ then the relations (1.1.1) are satisfied. But then clearly there exists a line bundle L with a local system having the c_{ij} as transition functions. One has $\kappa[L] = [a]$ proving surjectivity.

Let L^1, L^2 be any two line bundles. There exists a contractible locally finite covering $\mathcal{U} = \{U_i\}$, $i \in I$ and a local system $\{(U_i, s_i^1)\}$ and $\{(U_i, s_i^2)\}$ for L^1 and L^2 respectively. Let c_{ij}^1 and c_{ij}^2 be, respectively, the corresponding transition functions. Now if $\kappa[L^1] = \kappa[L^2]$ then we may choose logarithms $f_{ij}^k = \frac{1}{2\pi i} \log c_{ij}^k$, $k = 1,2$ so that $f_{ij} = f_{ij}^1 - f_{ij}^2$ satisfies $f_{ij} + f_{jk} - f_{ik} = 0$. But then if $\beta_i \in C(U_i)$ is defined by

$$\beta_i = \sum_{k \in I} f_{ki} h_k$$

$L^1 \to L^2 : s_i^1 \mapsto \lambda_i s_i^2$

one has

$$\beta_j - \beta_i = f_{ij} \quad .$$

Putting $\lambda_i = \exp 2\pi i \beta_i$ yields the relation $\lambda_i c_{ij}^1 \lambda_j^{-1} = c_{ij}^2$ proving $[L^1] = [L^2]$ or the injectivity of κ. That is

Proposition 1.2.1. <u>The map</u> $\kappa : \mathcal{L} \longrightarrow H^2(X, \underline{\underline{Z}})$ <u>is bijective.</u>

Remark 1.2.1. A line bundle L over X is called trivial if it is equivalent to the product bundle $\underline{\underline{C}} \times X$. That is, if and only if $S(X,L)$ contains a nowhere vanishing section. Clearly from the isomorphism (1.2.3) this is the case if and only if $\kappa[L] = 0$.

Since clearly $H^2(X,\underline{\underline{Z}}) = 0$ if X is contractible, it follows that if X is arbitrary and $\mathcal{U} = \{U_i\}$, $i \in I$, is any contractible covering of X and L is any line bundle over X, we may find a local system $\{(U_i,s_i)\}$, $i \in I$, for L. This follows since the restriction $L|U_i$ is the trivial bundle over U_i , and hence there is a nowhere vanishing smooth section s_i of L over U_i . Thus all line bundles over X have transition functions defined for a single fixed contractible covering.

1.3. For any manifold X let $\underline{\mathfrak{N}} = \underline{\mathfrak{N}}(X)$ denote the Lie algebra of all smooth complex vector fields on X and let $\Omega = \Omega(X)$ be the graded algebra of all complex-valued smooth differential forms on X.

If $\xi \in \underline{\mathfrak{N}}$ then $i(\xi)$ and $\Theta(\xi)$, respectively, denote interior product and Lie differentiation of Ω by ξ . Let d denote the exterior differentiation of Ω . The 3 operators are related by

$$\Theta(\xi) = i(\xi)\, d + d\, i(\xi) \ . \qquad \mathcal{L}_\xi \omega = \xi \lrcorner d\omega + d(\xi \lrcorner \omega) \tag{1.3.1}$$

where $i(\xi)$ lowers the degree (on Ω) by 1, $\Theta(\xi)$ preserves degrees and d raises degrees by 1.

One also has the relation, for ξ, $\eta \in \underline{\mathfrak{N}}$,

$$[\Theta(\xi), i(\eta)] = i([\xi,\eta]) \ . \qquad [\mathcal{L}_\xi \omega, \eta \lrcorner \omega] = [\xi,\eta] \lrcorner \omega \tag{1.3.2}$$

For a function $\phi \in C$ one has $i(\xi)\phi = 0$, $\Theta(\xi)\phi = \xi\phi$ and

$d\phi \in \Omega^1$ is given by

$$\langle d\phi, \xi \rangle = \xi\phi \; . \qquad (1.3.3)$$

If $\alpha \in \Omega^1$ one has $i(\xi)\alpha \in C$ is given by

$$i(\xi)\alpha = \langle \alpha, \xi \rangle \qquad (1.3.4)$$

$\Theta(\xi)\alpha$ is the 1-form given by

$$\langle \Theta(\xi)\alpha \;, \eta \rangle = \xi\langle \alpha, \eta \rangle - \; \alpha, \langle [\xi, \eta] \rangle \qquad (1.3.5)$$

and $d\alpha$ is the 2-form such that

$$d\alpha(\xi,\eta) = \xi\langle \alpha,\eta \rangle - \eta\langle \alpha,\xi \rangle - \langle \alpha,[\xi,\eta] \rangle \; . \qquad (1.3.6)$$

If $\omega \in \Omega^2$ then $i(\xi)\omega$ is the 1-form such that

$$\langle i(\xi)\omega, \eta \rangle = \omega(\xi,\eta) \quad . \qquad (1.3.7)$$

The condition that ω be closed, that is $d\omega = 0$, is that for $\xi,\eta,\zeta \in \underline{\mathfrak{V}}$ one has

$$\underset{\xi,\eta,\zeta}{\text{cyclic sum}}\; \xi\,\omega(\eta,\zeta) = \underset{\xi,\eta,\zeta}{\text{cyclic sum}}\; \omega([\xi,\eta],\zeta) \; . \qquad (1.3.8)$$

1.4. Now let L be any line bundle over X. A connection in L is a linear map

$$\nabla : \mathfrak{V} \longrightarrow \text{End } S, \quad \text{where} \quad \xi \longmapsto \nabla_\xi \; ,$$

such that for any $\phi \in C$ one has

$$(1) \quad \nabla_{\phi\xi} = \phi\nabla_{\xi} \qquad (1.4.1)$$

and for any $s \in S$

$$(2) \quad \nabla_{\xi} \phi s = (\xi\phi)\, s + \phi\nabla_{\xi} s \, . \qquad (1.4.2)$$

It is obvious from (1) that $(\nabla_{\xi} s)(p)$ for $p \in X$ depends only on s and the tangent vector ξ_p so that $\nabla_v s \in L_p$ is defined for any $s \in S$, $p \in X$, and a tangent vector v at p and $\nabla_{\xi_p} s = (\nabla_{\xi} s)(p)$ for any $\xi \in \underline{\mathfrak{w}}$.

Moreover from (2) it follows that $(\nabla_{\xi} s)(p)$ depends only on the germ of s at p. Thus if U is any open set, the connection ∇ in L induces a connection in $L|U$ and hence $\nabla_{\xi} s \in S(U) = S(U,L|U)$ is defined for all $\xi \in \underline{\mathfrak{n}}$ or $\underline{\mathfrak{n}}(U)$ and $s \in S(U)$.

If $x,y \in L_p$ where $x \neq 0$ we let $\frac{y}{x} \in \underline{\underline{C}}$ be that unique number c such that $y = cx$. More generally, if M is a manifold and $r,s: M \longrightarrow L^*$ are maps such that $\pi \circ r = \pi \circ s$ and such that s is nowhere vanishing we let $\frac{r}{s}$ be that function φ on M such that $r = \varphi s$. Clearly $\frac{r}{s} \in C(M)$ in case r and s are smooth.

Now assume that (L,∇) is a line bundle with connection over X. We now observe that if $U \subseteq X$ is any open set and $s \in S(U)$ is a nowhere vanishing section on U then one may associate to s a 1-form $\alpha(s) \in \Omega^1(U)$ as follows: If $t \in S(U)$ is arbitrary then $t/s \in C(U)$. But now the map $\underline{\mathfrak{n}} \longrightarrow C(U)$ defined by

$$\xi \longrightarrow \frac{1}{2\pi i} \frac{\nabla_\xi s}{s}$$

is clearly C-linear and hence there exists a unique 1-form $\alpha = \alpha(s) \in \Omega^1(U)$ such that for all $\xi \in \underline{\mathfrak{n}}$ one has

$$\nabla_\xi s = 2\pi i \langle \alpha, \xi \rangle s \tag{1.4.3}$$

where of course $\langle \alpha, \xi \rangle \in C(U)$ is short for $\langle \alpha, \xi | U \rangle$.

Let $S^*(U)$ be the set of all nowhere vanishing smooth sections over U. It is clearly a module for the group $C^*(U)$ of all nowhere vanishing functions on U.

Proposition 1.4.1. _Let_ (L, ∇) _be a line bundle with connection over_ X. _Let_ $U \subseteq X$ _be open and let_ $s, t \in S^*(U)$, _then one has_

$$\alpha(t) = \alpha(s) + \frac{1}{2\pi i} \frac{dg}{g} \tag{1.4.4}$$

where $g = \frac{t}{s} \in C^*(U)$.

Proof. For any $\xi \in \underline{\mathfrak{n}}$ one has

$$2\pi i \langle \alpha(t), \xi \rangle = \frac{\nabla_\xi t}{t} = \frac{\nabla_\xi gs}{gs} = \frac{\xi g}{g} + \frac{\nabla_\xi s}{s}$$

$$= \left\langle \frac{dg}{g} + 2\pi i \alpha(s), \xi \right\rangle .$$

This yields the relation (1.4.4) since ξ is arbitrary. QED

Corollary 1 to Proposition 1.4.1 _Let_ $\{(U_i, s_i)\}$, $i \in I$, _be a local system for the line bundle_ L _and let_ c_{ij} _be the corresponding transition functions_.

Now if ∇ _is a connection in_ L _and_ $\alpha_i = \alpha(s_i) \in \Omega^1(U_i)$ _then one has, in_ $U_i \cap U_j$

$$\alpha_j = \alpha_i + \frac{1}{2\pi i} \frac{d\, c_{ij}}{c_{ij}} \tag{1.4.5}$$

Conversely if a family of 1-forms $\alpha_i \in \Omega^1(U_i)$, $i \in I$, *are given satisfying* (1.4.5) *then there exists a unique connection* ∇ *such that* $\alpha(s_i) = \alpha_i$.

Proof. The first result follow immediately from (1.4.4) since in $U_i \cap U_j$ one has $c_{ij} = \frac{s_j}{s_i}$. On the other hand, if the 1-forms $\alpha_i \in \Omega^1(U_i)$ are prescribed arbitrarily satisfying (1.4.5) then $\nabla_\xi s \in S$ is well defined for all $\xi \in \underline{\mathfrak{n}}$ and $s \in S$ by putting in U_i

$$\nabla_\xi s = \left(\xi\left(\frac{s}{s_i}\right) + \frac{s}{s_i}\,(2\pi i \langle \alpha_i, \xi \rangle)\right) s_i \, . \tag{1.4.6}$$

The relation (1.4.6) guarantees that in $U_i \cap U_j$ the right side of (1.4.6) is unchanged if the index j replaces the index i . It is obvious that ∇ satisfies (1) and (2) of §1.4 so that it is a connection. It is furthermore clear that $\alpha(s_i) = \alpha_i$ and that having this property, ∇ is unique. QED

1.5 Now let L^* be the open subset of L given by the union $L^* = \bigcup L_p^*$ over all $p \in X$ where L_p^* is L_p minus the origin in L_p . Also let $\tilde{\pi} = \pi | L^*$. Then L^* is a bundle over X

$$\begin{array}{ccc} \underline{\underline{C}}^* & \longrightarrow & L^* \\ & & \Big\downarrow \tilde{\pi} \\ & & X \end{array}$$

with fiber $\underline{\underline{C}}^*$. In fact, by multiplication, $\underline{\underline{C}}^*$ obviously operates as a group of diffeomorphisms of L^* where the orbits are just the fibers L_p^* . The open submanifold L^* of L is of course the principal bundle associated to L .

Now the 1-form $\frac{1}{2\pi i}\frac{dz}{z}$ in $\underline{\underline{C}}^*$ is invariant under multiplication by $\underline{\underline{C}}^*$. It follows therefore that for any $p \in X$ there exists a unique 1-form β_p in L_p^* such that for any $\underline{\underline{C}}^*$ - map

$$\tau : \underline{\underline{C}}^* \longrightarrow L_p^*$$

one has $\tau^*(\beta_p) = \frac{1}{2\pi i}\frac{dz}{z}$.

Now a connection form in L^* is by definition a 1-form $\alpha \in \Omega^1(L^*)$ such that

(1) α is invariant under $\underline{\underline{C}}^*$

(2) For all $p \in X$ one has $\alpha | L_p^* = \beta_p$.

If M is an arbitrary manifold and $s: M \longrightarrow L^*$ we let

$$\sigma_s : \underline{\underline{C}}^* \times M \longrightarrow L^* \tag{1.5.1}$$

be the map given by $\sigma_s(c,p) = c\, s(p)$.

Lemma 1.5.1. *Assume* α *is a connection form in* L^* *and let* s *be as above and smooth. Then*

$$\sigma_s^*\alpha = \left(\frac{1}{2\pi i}\frac{dz}{z}, s^*\alpha\right) . \tag{1.5.2}$$

Moreover if $r: M \longrightarrow L^*$ *is any other smooth map such that* $\pi \circ s = \pi \circ r$ *so that* $g = \frac{r}{s} \in C^*(U)$ *then*

$$r^*\alpha = s^*\alpha + \frac{1}{2\pi i}\frac{dg}{g} . \tag{1.5.3}$$

Proof. For any $p \in M$ it is clear from the definition of α that $\sigma_s^*\alpha | \underline{\underline{C}}^* \times p = \left(\frac{1}{2\pi i}\frac{dz}{z}, 0\right)$. However σ_s is a $\underline{\underline{C}}^*$-map (where $\underline{\underline{C}}^*$ operates on $\underline{\underline{C}}^* \times M$ in the obvious way) so that $\sigma_s^*\alpha$ is $\underline{\underline{C}}^*$-invariant and hence $\sigma_s^*\alpha$ is determined by $\sigma_s^*\alpha|1 \times M = (0, s^*\alpha)$. This yields (1.5.2) .

Now let $\rho : \underline{\underline{C}}^* \times M \longrightarrow C^* \times M$ be the map given by $\rho(c,p) = (g(p)c, p)$. Clearly

$$\rho^* \sigma_s^*\alpha = \sigma_s^*\alpha + \left(0, \frac{1}{2\pi i}\frac{dg}{g}\right). \tag{1.5.4}$$

But $\sigma_r = \sigma_s \circ \rho$ so that $\sigma_r^*\alpha = \rho^*\sigma_s^*\alpha$. But $\sigma_r^*\alpha = \frac{1}{2\pi i}\left(\frac{dz}{z}, r^*\alpha\right)$ by (1.5.2). Thus (1.5.3) follows from (1.5.4). QED

Proposition 1.5.1 Let X be a manifold and let L be a line bundle over X . Then if ∇ is a connection in L there exists a unique connection form $\alpha \in \Omega^1(L^*)$ such that for all open $U \subseteq X$ and all $s \in S^*(U)$ $(s\colon U \longrightarrow L^*)$ one has

$$\alpha(s) = s^*(\alpha) . \tag{1.5.5}$$

Conversely if α is a connection form in L^* then there is a unique connection ∇ in L such that (1.5.5) is satisfied for all $s \in S^*(U)$ and all open $U \subseteq X$.

Proof. Now for any $s \in S^*(U)$ clearly

$$\sigma_s : \underline{\underline{C}}^* \times U \longrightarrow \tilde{\pi}^{-1}(U)$$

is a diffeomorphism where σ_s is given by (1.5.1).

Now assume α is a connection form in L^*. But now it $t \in S^*(U)$ is arbitrary then, by Lemma 1.5.1,

$$t^*(\alpha) = s^*(\alpha) + \frac{1}{2\pi i}\frac{dg}{g} .$$

Given a local system (U_i, s_i), $i \in I$, one therefore has in $U_i \cap U_j$, $s_j^*(\alpha) = s_i^*(\alpha) + \frac{1}{2\pi i}\frac{dc_{ij}}{c_{ij}}$.

By Corollary 1 to Proposition 1.4.1. there exists a unique connection ∇ such that $\alpha(s_i) = s_i^*(\alpha)$ for all i. One has $\alpha(s) = s^*(\alpha)$ for any $s \in S^*(U)$ and any open $U \subseteq X$ by (1.4.4) and (1.5.3).

Thus a connection form α in L uniquely determines a connection ∇ in L such that (1.5.5) is satisfied.

Conversely, assume ∇ is given. For $s \in S^*(U)$ let β_s be the 1-form in $\underline{\underline{C}}^* \times U$ given by $\beta_s = \left(\frac{1}{2\pi i}\frac{dz}{z}, \alpha(s)\right)$ and let α_s be the 1-form in $\tilde{\pi}^{-1}(U)$ defined so that $\beta_s = \sigma_s^*(\alpha_s)$. Clearly α_s is a connection form in $L^*|U$. We assert that for any $t \in S^*(U)$ one has

$$\alpha_s = \alpha_t . \tag{1.5.6}$$

Indeed if $g = \frac{t}{s}$ then where ρ is as in (1.5.4)

$$\sigma_t^*(\alpha_s) = \rho_g^*(\sigma_s^*(\alpha_s)) = \rho_g^*(\beta_s) = \left(\frac{1}{2\pi i}\frac{dz}{z}, \alpha(s) + \frac{1}{2\pi i}\frac{dg}{g}\right) .$$

But $\alpha(s) + \frac{1}{2\pi i}\frac{dg}{g} = \alpha(t)$ by (1.4.4). Thus $\sigma_t^*(\alpha_s) = \beta_t$.

But $\sigma_t^*(\alpha_t) = \beta_t$. This proves (1.5.6). But (1.5.6) clearly implies, using for example a local system, there exists a connection form α in L^* such that for any open set $U \subseteq X$ and any $s \in S^*(U)$ one has $\sigma_s^*(\alpha) = \left(\frac{1}{2\pi i}\ \frac{dz}{z}\ ,\ \alpha(s)\right)$. But by (1.5.2) one has $\alpha(s) = s^*(\alpha)$. The connection form α is clearly unique having the property (1.5.5) since by (1.5.2) α is determined by knowing $s^*(\alpha)$ for all possible s . QED

1.6. Henceforth a line bundle with connection ∇ will be denoted by the pair (L,α) where α is the connection form in L^* corresponding to ∇ .

Proposition 1.6.1. <u>Let</u> (L,α) <u>be a line bundle with connection over</u> X . <u>Then there exists a unique closed</u> 2-<u>form</u> $\omega \in \Omega^2(X)$ <u>such that</u>

$$d\alpha = (\tilde{\pi})^*\omega \ . \tag{1.6.1}$$

<u>Moreover if</u> $U \subseteq X$ <u>is open and</u> $s \in S^*(U)$ <u>is arbitrary then</u>

$$d\alpha(s) = \omega | U \ . \tag{1.6.2}$$

<u>Proof</u>. If $s \in S^*(U)$ then by (1.5.2) one has $\sigma_s^*\alpha = \left(\frac{1}{2\pi i}\ \frac{dz}{z}\ ,\alpha(s)\right)$. But then $d(\sigma_s^*(\alpha)) = (0, d\alpha(s))$. However $d(\sigma_s^*\alpha) = \sigma_s^*\, d\alpha$. Thus

$$d\alpha | \tilde{\pi}^{-1}(U) = \tilde{\pi}^*\ d(\alpha(s)\ ) \ . \tag{1.6.3}$$

But now if $t \in S^*(U)$ then $d(\alpha(t)) = d(\alpha(s))$ by (1.4.4) since $\frac{dg}{g}$ is closed. Hence using a local system there exists a unique closed 2-form ω on X such that $d(\alpha(s)) = \omega | U$ for any $s \in S^*(U)$ where

$U \subseteq X$ is open. By (1.6.3) one has $d\alpha = \tilde{\pi}^* \omega$. The 2-form ω is uniquely characterized by this relation since $\tilde{\pi}^*$ maps $\Omega(X)$ injectively into $\Omega(L^*)$ (using for example the local product structure in L^*). QED

The closed 2-form ω is called the curvature of (L,α) and is written

$$\omega = \operatorname{curv}(L,\alpha) .$$

1.7. If M is a manifold and $p \in M$ then $T_p(M)$ denotes the tangent space to M at P .

A curve γ in a manifold M is a continuous map $\gamma: I \longrightarrow M$ where $I \subseteq \underline{\underline{R}}$ is an interval. It is called piece-wise smooth if I is a finite union of intervals on each of which γ is smooth. If γ is a smooth curve, then $\gamma'(t) \in T_{\gamma(t)}(M)$ will denote the tangent vector to γ corresponding to $t \in I$.

If L is a line bundle over X and $\gamma: I \longrightarrow X$ is a curve in X , then a section along γ is a curve $r: I \longrightarrow L$ in L such that $r(t) \in L_{\gamma(t)}$ for all $t \in I$. If α is a connection form in L^* we wish to recall the notion of auto-parallel sections along piece-wise smooth curves in X .

Assume the image of a curve $\gamma: I \longrightarrow X$ lies in an open set $U \subseteq X$ such that $S^*(U)$ is not empty. Then, if $s \in S^*(U)$, $s \circ \gamma$ is a nowhere vanishing section along γ and hence $\frac{r}{s \circ \gamma}$ is a complex-valued function on I whenever r is a section along γ . If r (and hence γ, by projection) is smooth it is clear that $\phi \in C(I)$.

If r is smooth one defines the covariant derivative of r to be the (smooth) section ∇r along γ given by

$$\left(\frac{\nabla r}{s\circ\gamma}\right)(t) = \left(\frac{d}{dt}\,\frac{r}{s\circ\gamma}\right)(t) + 2\pi i\left(\frac{r}{s\circ\gamma}\right)(t)\langle\alpha(s),\gamma'(t)\rangle . \tag{1.7.1}$$

This definition of ∇r is easily seen to be independent of the choice of U or of $s \in S^*(U)$ (using the relation (1.4.4)) and hence ∇r is defined for all smooth r, making no assumption about the image of γ.

Remark 1.7.1. Note that if r is of the form $r = s \circ \gamma$ where $s \in S$ then $(\nabla r)(t)$ may be given by

$$(\nabla r)(t) = \nabla_{\gamma'(t)}\, s . \tag{1.7.2}$$

If r is a nowhere vanishing smooth section along a curve γ then ∇r can be described in a more direct way. Indeed $r: I \longrightarrow L^*$ so that $r'(t)$ is a tangent vector to L^* at $r(t)$ and hence $\langle\alpha, r'\rangle$ is a smooth function on I.

Proposition 1.7.1. If r is nowhere vanishing smooth section along γ one has

$$\nabla r = 2\pi i\langle\alpha, r'\rangle r . \tag{1.7.3}$$

Proof. Since the statement is local we can assume $\gamma(I) \subseteq U$ using the notation above. But

$$\langle\alpha(s), \gamma'\rangle = \langle s^*\alpha, \gamma'\rangle = \langle\gamma^* s^*\alpha, \frac{d}{dt}\rangle = \langle(s\circ\gamma)^*\alpha, \frac{d}{dt}\rangle .$$

But by (1.5.3)

$$r^*\alpha = (s \circ \gamma)^*\alpha + \frac{1}{2\pi i}\, d\left(\frac{r}{s\circ\gamma}\right) \frac{s\circ\gamma}{r} .$$

Hence $2\pi i \langle\alpha, r'\rangle = 2\pi i \langle r^*\alpha, \frac{d}{dt}\rangle = 2\pi i \langle\alpha(s), \gamma'\rangle + \frac{d}{dt}\left(\frac{r}{s\circ\gamma}\right)\frac{s\circ\gamma}{r}$.

But the right side equals $\left(\frac{\nabla r}{s\circ\gamma}\right) \frac{s\circ\gamma}{r} = \frac{\nabla r}{r}$ by (1.7.1).

Hence $\nabla r = 2\pi i \langle\alpha, r'\rangle r$. QED

Remark 1.7.2. If $\gamma: I \longrightarrow X$ then the easily verified relation

$$\nabla f r = f' r + f \nabla r . \tag{1.7.4}$$

for $f \in C(I)$ and r a smooth section along γ together with (1.7.2) gives a convenient description of ∇s for any smooth section s along γ .

1.8. A section r along a curve γ in X is called auto-parallel if it (and hence γ) is piece-wise smooth and $\nabla r = 0$ along those intervals on which it is smooth.

Now assume $\gamma: I \longrightarrow X$ is a curve in X when $\gamma(I) \subseteq U$ and $s \in S^*(U)$. Assume r is a smooth section along γ . The condition that $\nabla r = 0$ is, by (1.7.1), that

$$\frac{d}{dt}\left(\frac{r}{s\circ\gamma}\right) = -2\pi i \frac{r}{s\circ\gamma} \langle\alpha(s), \gamma'\rangle .$$

But then if $t_o \in I$ is fixed $\left(\frac{r}{s\circ\gamma}\right)(t)$ is uniquely determined by $\left(\frac{r}{s\circ\gamma}\right)(t_o)$ (which may be arbitrary) and is given by the line integral of α

$$\frac{r}{s\circ\gamma}(t) = \frac{r}{s\circ\gamma}(t_o)\, e^{-2\pi i \int_{\gamma_t} \alpha(s)} \tag{1.8.1}$$

where γ_t is the curve $\gamma|[t_o,t]$ (oriented in going from t_o to t). It follows then that if γ is any piece-wise smooth curve in X and $x \in L_{\gamma(t)}$ there exists a unique autoparallel section r_x along γ such that $r_x(t) = x$. Hence if $\gamma:[a,b] \longrightarrow X$ we can define a linear isomorphism

$$P_\gamma : L_{\gamma(a)} \longrightarrow L_{\gamma(b)}$$

called parallel transport along γ by the condition that $P_\gamma(x) = r_x(b)$ where $x = r_x(a) \in L_{\gamma(a)}$.

Remark (1.8.1). Of course (1.8.1) is valid for a piecewise smooth curve γ rather than just a smooth curve provided only that γ lies in U .

Now let $\Gamma = \Gamma(X)$ be the set of all piecewise smooth closed curves on X . Consider the function (the scalar parallel transport function)

$$Q : \Gamma \longrightarrow \underline{\underline{C}}^*$$

defined as follows: Any $\gamma \in \Gamma$ has the same initial and end point p . Thus $P_\gamma : L_p \longrightarrow L_p$ and hence there exists a non-zero complex number $Q(\gamma)$ such that for any $x \in L_p$

$$P_\gamma(x) = Q(\gamma)\, x \ . \tag{1.8.2}$$

We recall that $\gamma \in \Gamma$, where $\gamma : I \longrightarrow X$, is called homotopic to a point if there is a "rectangle" $R = [a,b] \times [c,d]$ in the plane and a continuous map $\sigma : R \longrightarrow X$ together with a piecewise smooth parameterization $\rho : I \longrightarrow \dot{R}$ of the boundary $\dot{R}$ of R oriented counterclockwise such that $\sigma \circ \rho = \gamma$. In such a case one knows that we may choose σ so that $[a,b]$ and $[c,d]$ may be divided into intervals I_i and J_j with the property that $\sigma_{ij} = \sigma | R_{ij}$ where $R_{ij} = I_i \times J_j$ is smooth. Such a map σ defines an oriented surface (with $\dot{R}$ oriented couterclockwise as boundary) and is called a surface of deformation of γ.

Theorem 1.8.1. _Let_ (L,α) _be a line bundle with connection over a manifold_ X. _Let_ γ _be a closed piece-wise smooth curve in_ X _which is homotopic to a point and let_ σ _be any surface of deformation of_ γ. _Let_ $\omega = \text{curv}\,(L,\alpha)$. _Then the scalar multiplication_ $Q(\gamma)$, _induced by parallel transport around_ γ, _is given by the surface integral_

$$Q(\gamma) = e^{-2\pi i \int_\sigma \omega} \tag{1.8.3}$$

Proof. It is clear that we can choose the R_{ij} sufficiently small so that there exists an open set $U_{ij} \subseteq X$ with an element $s_{ij} \in S^*(U_{ij})$ and such that $\sigma(R_{ij}) \subseteq U_{ij}$. But the boundary of R_{ij}, oriented counterclockwise, defines a closed curve γ_{ij} in U_{ij} and by (1.8.1) (see Remark 1.8.1) one has

$$Q(\gamma_{ij}) = e^{-2\pi i \int_{\gamma_{ij}} \alpha(s_{ij})} .$$

But by Stokes' theorem this is given in invariant form by the surface integral

$$Q(\gamma_{ij}) = e^{-2\pi i \int_{\sigma_{ij}} \omega}$$

since $\omega | U_{ij} = d\alpha(s_{ij})$ by (1.6.2). However it is clear that $Q(\gamma)$ is a product of $Q(\gamma_{ij})$ over all i and j and hence one obtains (1.8.3). QED

1.9. Let L be a line bundle over X. A Hermitian structure in L is function H on the set of all $(x,y) \in L \times L$ where $\pi(x) = \pi(y)$ such that (1) H induces a 1-dimensional Hilbert space structure on L_p for all $p \in X$ and (2) one has $|H|^2 \in \underline{\underline{C}}^*(L^*)$ where $|H|^2$ is the positive-valued function on L^* defined by $|H|^2 (x) = H(x,x)$.

When H is understood we write (x,y) for $H(x,y)$ when $x,y \in L_p$ and (s,t) for the (clearly measureable) function on X defined by $(s,t)(p) = (s(p), t(p))$ for $s,t \in S_m$. Also we let $|s|^2 = (s,s)$. Obviously $(s,t) \in C(X)$ if $s,t \in S$ and $|s|^2 \in C^*(U)$ if $s \in S^*(U)$.

If α is a connection form in L^* then H will be called α-invariant if

$$\xi(s,t) = (\nabla_\xi s, t) + (s, \nabla_\xi t) \tag{1.9.1}$$

for all $s,t \in S$ and all real $\xi \in \underline{\mathfrak{U}}$.

Proposition 1.9.1. <u>Let</u> (L,α) <u>be given. Then a necessary and sufficient condition that</u> L <u>possess an</u> α-<u>invariant Hermitian</u>

<u>structure is that the real 1-form</u> $2\pi i\,(\alpha-\bar{\alpha})$ <u>on</u> L^* <u>be exact</u>. <u>(The bar denotes the conjugate form.)</u>

<u>Moreover in such a case</u> H <u>is unique up to a positive constant and one has</u>

$$2\pi i\,(\alpha-\bar{\alpha}) = \frac{d|H|^2}{|H|^2} = d \log |H|^2 \quad . \tag{1.9.2}$$

<u>Proof</u>. Assume H is an α-invariant Hermitian structure in L . Then if $U \subseteq X$, $s \in S^*(U)$ and $\xi \in \mathfrak{n}$ is real, one has

$$\begin{aligned} \xi|s|^2 &= \xi(s,s) = (\nabla_\xi s,\, s) + (s,\, \nabla_\xi s) \\ &= 2\pi i\,(\alpha(s),\xi)\,|s|^2 - 2\pi i\,(\bar{\alpha}(s),\xi)\,|s|^2 \end{aligned} \tag{1.9.3}$$

or

$$\left\langle \frac{d|s|^2}{|s|^2},\, \xi \right\rangle = \frac{\xi|s|^2}{|s|^2} = 2\pi i \langle (\alpha(s) - \bar{\alpha}(s))\,,\, \xi \rangle .$$

This proves that

$$\frac{d|s|^2}{|s|^2} = 2\pi i\,\{\alpha(s) - \bar{\alpha}(s)\} \quad . \tag{1.9.4}$$

On the other hand using the notation of § 1.5 one has that

$$\sigma_s^*\alpha = \left(\frac{1}{2\pi i}\,\frac{dz}{z}\,,\, \alpha(s)\right)$$

in $\underline{\underline{C}}^* \times U$. Hence

$$\begin{aligned} \sigma_s^*(2\pi i\,(\alpha-\bar{\alpha})) &= \left(\frac{d|z|^2}{|z|^2}\,,\, \frac{d|s|^2}{|s|^2}\right) \\ &= \sigma_s^*\left(\frac{d|H|^2}{|H|^2}\right) \end{aligned}$$

since $|H|^2 (\sigma_s(z,p)) = |z|^2\,|s(p)|^2$.

Thus $2\pi i\,(\alpha-\bar{\alpha}) = \dfrac{d|H|^2}{|H|^2}$ in $\tilde{\pi}^{-1}(U)$ and hence everywhere in L^*.

This proves $2\pi i\,(\alpha-\bar{\alpha})$ is exact and H is unique up to a positive scalar multiple. (Indeed since $2\pi i\,(\alpha-\bar{\alpha}) = d\log|H|^2$, it follows that $\log|H|^2$ is unique up to an additive constant.

Now conversely assume $2\pi i\,(\alpha-\bar{\alpha})$ is exact. Then since it is real there is a real $g \in C(L^*)$ such that $2\pi i\,(\alpha-\bar{\alpha}) = dg$ and hence if $h = e^g$ one has $h > 0$ and $2\pi i\,(\alpha-\bar{\alpha}) = \dfrac{dh}{h}$. But then (1.5.2) implies

$$\sigma_s^*\left(\frac{dh}{h}\right) = \left(\frac{d|z|^2}{|z|^2},\ 2\pi i\left(\alpha(s) - \overline{\alpha(s)}\right)\right) \tag{1.9.5}$$

for any open $U \subseteq X$ and $s \in S^*(U)$. But from the left factor on the right side one has $h(zx) = |z|^2 h(x)$ for any $z \in \underline{\underline{C}}^*$ and $x \in L^*$ so that there exists a unique Hermitian structure H such that $|H|^2 = h$. However (1.9.5) then implies (1.9.4) since $h \circ s = |s|^2$. But (1.9.4) implies (1.9.3) using (1.4.3) and using (1.4.2) one obtains (1.9.1). QED

Corollary 1 to Proposition 1.9.1. <u>If</u> L <u>possesses</u> <u>an</u> α-<u>invariant</u> <u>Hermitian</u> <u>structure</u> <u>then</u> $\omega = \text{curv}\,(L,\alpha)$ <u>is</u> <u>a</u> <u>real</u> 2-<u>form</u>.

<u>Proof</u>. Since $\tilde{\pi}^*\omega = d\alpha$ one also has $\tilde{\pi}^*\omega = d\bar{\alpha}$ because $\alpha - \bar{\alpha}$ is exact and hence closed. Thus $\omega = \bar{\omega}$ since $\tilde{\pi}^*\bar{\omega} = d\bar{\alpha}$. QED

One may express the existence of an α-invariant Hermitian structure in terms of the scalar parallel function Q defined in § 1.8. We need

Lemma 1.9.1. *Let* (L,α) *be a line bundle with connection over* X. *Let* $p \in X$ *and let* $U \subseteq X$ *be any coordinate neighborhood with coordinates* $u^1,\ldots,u^m \in C(U)$ *such that* $q \longrightarrow \left(u^1(q)\ldots u^m(q)\right)$ *is a diffeomorphism of* U *with the open unit ball in* $\underline{\underline{R}}^m$. *Choose* U *small enough so that there exists* $s \in S^*(U)$. *For each* $q \in U$ *let* γ_q *be the radial curve joining* p *to* q *so that*

$$P_{\gamma_q} : L_p \longrightarrow L_q$$

is a linear isomorphism. Let w *be the section on* U *defined by* $u(q) = P_{\gamma_q}\left(s\,(p)\right)$ *then* w *is smooth (i.e.* $w \in S^*(U)$ *) and in fact*

$$\frac{w}{s}(q) = e^{-2\pi i \int_{\gamma_q} \alpha(s)} . \qquad (1.9.6)$$

Proof. This is an immediate consequence of (1.8.1) together with the obvious fact that the map $q \longrightarrow e^{-2\pi i \int_{\gamma_q} \alpha(s)}$ is a smooth non-vanishing function on U. QED

Proposition 1.9.2. *Let* (L,α) *be given. Then* L *possesses an* α*-invariant Hermitian structure* H *if and only if the image of the scalar parallel transport function*

$$Q : \Gamma \longrightarrow \underline{\underline{C}}^*$$

(*see* (1.8.2)) *lies on the unit circle* $\underline{\underline{T}} \subseteq \underline{\underline{C}}^*$.

Moreover if H *is an arbitrary Hermitian structure in* L *then* H *is* α*-invariant if and only if for any piecewise smooth curve* $\gamma:[a,b] \longrightarrow X$ *the map* $P_\gamma : L_{\gamma(a)} \longrightarrow L_{\gamma(b)}$ *is an isometry*.

<u>Proof</u>. Assume H is a Hermitian structure and $\gamma:[a,b]\longrightarrow X$ is a smooth curve. Then clearly P_γ is an isometry if and only if $\frac{d}{dt}\,|r(t)|^2 = 0$ where r is a non-zero auto-parallel section along γ. This may be expressed by the relation

$$\frac{d}{dt}\,(r,r) = (\nabla r,r) + (r,\nabla r)\ . \tag{1.9.7}$$

But since one clearly has $\nabla(\phi u) = \left(\frac{d}{dt}\,\phi\right)\,u + \phi\,\nabla u$ for any $\phi \in C[a,b]$ and u a smooth section along γ, it follows that (1.9.7) holds if and only if for any $t_o \in [a,b]$ and $s \in S^*(U)$ where $\gamma(t_o) \in U$ one has

$$\frac{d}{dt}\,|u|^2 = (\nabla u,u) + (u,\nabla u)$$

where u is the section along the restriction of γ to a neighborhood t_o defined by $u(t) = s(\gamma(t))$. But $\frac{d}{dt}\,|u|^2\,(t) = \gamma'(t)\,|s|^2$ and $(\nabla u)(t) = \nabla_{\gamma'(t)}s$. Since any real tangent vector is the tangent to some smooth curve one has that (1.9.7) holds if and only if (1.9.3) holds for all real $\xi \in \underline{\mathfrak{n}}$ and all $s \in S^*(U)$ for all U. But by (1.4.2) this holds if and only if H is α-invariant.

This proves the second statement of the Proposition. It also proves that Q is $\underline{\underline{T}}$-valued in case H is an α-invariant Hermitian structure.

Now finally assume that (L,α) is given and Q is $\underline{\underline{T}}$-valued. We show the existence of an α-invariant Hermitian structure. Let $p \in X$ be fixed and let H_o be a fixed Hilbert space structure in the line L_o. Now let $p \in X$ be arbitrary and let γ be any piece-wise smooth curve in X joining o to p. We let H_p be the unique Hilbert space structure in L_p defined so that $P_\gamma : L_o \longrightarrow L_p$ is an isometry. We now observe that H_p is independent of γ since Q, defined on a closed curve, is $\underline{\underline{T}}$-valued. This clearly defines H which is a Hermitian structure in case $|H|^2$ is smooth. To prove smoothness we have only to show that for any $p \in X$ there exists a neighborhood U of p and a (smooth) section $w \in S^*(U)$ such that $|w|^2$ is smooth. But this follows using the section w given in Lemma 1.9.1. By Lemma 1.9.1 $w \in S^*(U)$ but, by definition, clearly $|w|^2$ is constant in U.

But now H is α-invariant since by construction P_γ is an isometry for all piece-wise smooth curves γ in X. QED

1.10 Let (L^1, α^i), $i = 1,2$, be two line bundles with connection over X. We say that they are equivalent if there exists a line bundle equivalence

$$\tau : L^1 \longrightarrow L^2$$

such that $\tau^*(\alpha^2) = \alpha^1$.

Remark 1.10.1. It is immediate that if τ is a line bundle equivalence then τ defines an equivalence of line bundles with connection if and only if for all $\xi \in \underline{\mathcal{U}}$ and $s \in S(X,L^1)$ one has

$$\tau(\nabla^1_\xi s) = \nabla^2_\xi \tau s \tag{1.10.1}$$

where ∇^i, $i = 1,2$, are the respective connections in L^1 and L^2.

Indeed if $\tau^*(\alpha^2) = \alpha^1$ then for any open $U \subseteq X$ and $s \in S^*(U)$ one has $(\tau s)^*(\alpha^2) = s^*(\tau^*(\alpha^2)) = s^*(\alpha^1)$ which implies (1.10.1) by (1.4.3) for such an s and hence for all $t \in S$ using a local system. Reversing the argument, if (1.10.1) holds then $\tau^*(\alpha^2) = \alpha^1$ since $\tau^*(\alpha^2)$ is clearly a connection form in L^1. But (1.10.1) implies $s^*(\tau^*(\alpha^2)) = s^*(\alpha^1)$ for all $s \in S^*(U)$ over all open $U \subseteq X$. This implies $\tau^*(\alpha^2) = \alpha^1$ by the uniqueness part of Proposition 1.5.1.

Now if L is a line bundle over X and $\phi \in C^*(X)$ then ϕ defines an equivalence of L with itself

$$\tau_\phi : L \longrightarrow L \tag{1.10.2}$$

by putting $\tau_\phi x = \phi(\pi(x))\, x$ for any $x \in L$.

Remark 1.10.2. Note of course that any equivalence of L with itself is uniquely of this form.

Lemma 1.10.1. <u>Let</u> (L,α) <u>be given over</u> X <u>and let</u> $\phi \in C^*(X)$ <u>then</u>

$$(\tau_\phi)^*\alpha = \alpha + \frac{1}{2\pi i}\frac{d\tilde{\phi}}{\tilde{\phi}} \tag{1.10.3}$$

<u>where</u> $\tilde{\phi} \in C^*(L^*)$ <u>is just the pull back</u> $\phi \circ \tilde{\pi}$.

<u>Moreover if</u> α_1 <u>is another connection form in</u> L <u>then</u> (L,α) <u>and</u> (L,α_1) <u>are equivalent if and only if</u>

$$\alpha_1 = \alpha + \frac{1}{2\pi i}\frac{d\tilde{\phi}}{\tilde{\phi}}$$

<u>where</u> $\phi \in C^*(X)$.

Proof. Let $U \subseteq X$ be open. Let $s \in S^*(U)$ and let $\sigma_s : \underline{\underline{C}}^* \times U \longrightarrow \tilde{\pi}^{-1}(U)$ be defined as in (1.5.1). Then

$$\tau_\phi \circ \sigma_s = \sigma_t \qquad (1.10.4)$$

where $t \in S^*(U)$ is given by $t(p) = \phi(p)\, s(p)$. But then

$$\sigma_s^*(\tau_\phi^*\alpha) = \sigma_t^*\,(\alpha) = \left(\frac{1}{2\pi i}\,\frac{dz}{z}\,,\ \alpha(t)\right) .$$

But $\alpha(t) = \alpha(s) + \frac{1}{2\pi i}\,\frac{d\phi_o}{\phi_o}$ by (1.4.4) where $\phi_o = \phi|U$.

Hence $\sigma_s^*(\tau_\phi^*\alpha - \alpha) = \left(0,\ \frac{1}{2\pi i}\,\frac{d\phi_o}{\phi_o}\right) = \sigma_s^*\left(\frac{1}{2\pi i}\,\frac{d\tilde{\phi}}{\tilde{\phi}}\right)$.

Thus $\tau_\phi^*\alpha = \alpha + \frac{1}{2\pi i}\,\frac{d\tilde{\phi}}{\tilde{\phi}}$ on $\tilde{\pi}(U)$ (since σ_s is a diffeomorphism). This proves (1.10.3). The second statement of the Lemma follows immediately from Remark 1.10.1. QED

Corollary 1 to Lemma 1.10.1. Let $\tau: L \longrightarrow L$ be an equivalent of L with itself. Then τ is an equivalence of (L,α) with itself if and only if τ is global multiplication by a complex constant $c \in \underline{\underline{C}}^*$.

Proof. By Lemma 1.10.1 one has $\tau_\phi^*\alpha = \alpha$ if and only if $\frac{d\tilde{\phi}}{\tilde{\phi}} = 0$. This implies $\tilde{\phi}$ and hence ϕ is constant. QED

1.11. Let L_i, $i = 1,2$, be line bundles over, respectively, manifolds X_i, $i = 1,2$. A map of line bundles is a smooth map $\tau: L_1 \longrightarrow L_2$ such that there exists a smooth map $\check{\tau}: X_1 \longrightarrow X_2$ (necessarily unique) such that (1) the diagram

$$\begin{array}{ccc} L_1 & \xrightarrow{\tau} & L_2 \\ \downarrow{\scriptstyle \pi_1} & & \downarrow{\scriptstyle \pi_2} \\ X_1 & \xrightarrow{\check{\tau}} & X_2 \end{array}$$

is commutative and such that (2) for all $p \in X$, the map $\tau:(L_1)_p \to (L_2)_{\check{\tau}p}$ is a linear isomorphism.

A smooth map $\rho: X_1 \longrightarrow X_2$ is called liftable, relative to L_1 and L_2, if there exists a map (called a lifting) of line bundles $\tau: L_1 \longrightarrow L_2$ such that $\check{\tau} = \rho$.

Lemma 1.11.1. *Let* $\rho: X_1 \longrightarrow X_2$ *be a smooth map and let* L_1, L_1' *be line bundles over* X_1 *and* L_2 *a line bundle over* X_2 . *Then if* ρ *can be lifted to* $\tau: L_1 \longrightarrow L_2$ *it can be lifted to* $\tau': L_1' \longrightarrow L_2$ *if and only if* L_1 *and* L_1' *are equivalent and in such a case there is a unique equivalence* $\mu: L_1' \longrightarrow L_1$ *such that* $\tau \circ \mu = \tau'$.

Proof. It is clear that if $\mu: L_1' \longrightarrow L_1$ is an equivalence of line bundles then $\tau' = \tau \circ \mu$ is a lifting of ρ . On the other hand if $\tau': L_1' \longrightarrow L_2$ is a lifting of ρ it is clear that ignoring smoothness, there is a unique map of line bundles $\mu: L_1' \longrightarrow L_1$ such that (1) $\tau' = \tau \circ \mu$ and such that (2) $\check{\mu}$ is the identity on X_1 . It suffices only to prove that μ is a diffeomorphism.

Let $p_1 \in X_1$ and let $p_2 = \rho(p_1)$. Let U_2 be a neighborhood of p_2 with an element $s_2 \in S^*(U_2, L_2)$. Also let U_1 be a neighborhood of p_1 in X_1 such that $\rho(U_1) \subseteq U_2$ and such that one has elements $s_1 \in S^*(U_1, L_1)$ and $s_1' \in S^*(U_1, L_1')$. Let $\nu: \underline{\underline{C}}^* \times U_2 \longrightarrow \underline{\underline{C}}^*$

be the projection on the first component. Then in the notation of (1.5.1.) let $f, f' \in C^*(U_1)$ be given by $f = \nu \circ \sigma_{s_2}^{-1} \circ \tau \circ s_1$ and $f' = \nu \circ \sigma_{s_2}^{-1} \circ \tau' \circ s_1'$. But then $\mu | (\tilde{\pi}_1')^{-1}(U_1) = \sigma_{s_1} \circ g \circ \sigma_{s_1'}^{-1}$ where $g: \underline{\underline{C}}^* \times U_1 \longrightarrow \underline{\underline{C}}^* \times U_1$ is given by $g = 1 \times f / f'$ proving that μ is a diffeomorphism.

Remark 1.11.1. It is easy and well known that a map $\rho: X_1 \longrightarrow X_2$ induces a homomorphism $\rho^*: H^i(X_2, A) \longrightarrow H^i(X_1, A)$ for any i. Given L_1 and L_2 as above, one can easily show that ρ is liftable relative to L_1 and L_2 if and only if $\rho^* \kappa[L_2] = \kappa[L_1]$. (See Proposition 1.2.1).

Remark 1.11.2. If as above $\tau: L_1 \longrightarrow L_2$ is a map of line bundles then τ induces a linear map on the respective spaces of sections

$$\tau^*: S_2 \longrightarrow S_1$$

where $S_i = S(X_2, L_2)$ such that if $s \in S^2$ then $\tau^* s \in S_1$ is given by the relation

$$\tau\left((\tau^* s)(p_1)\right) = s\ (\check{\tau} p_1) \tag{1.11.1}$$

for any $p_1 \in X_1$. To see the smoothness of $\tau^* s$ note that in the notation of the proof of Lemma 1.11.1 one has in U_1

$$\tau^* s = \left(\frac{s}{s_2} \circ \rho\right) f^{-1}\ s_1 \ . \tag{1.11.2}$$

We extend the notion above to the case of line bundles with connection. Let X_1, X_2 be as above and let (L_i, α_i), $i = 1,2$ be line bundles with connection over X_i respectively. A map $\tau: L_1 \longrightarrow L_2$ of line bundles is called a map of line bundles with connection in case $\tau^* \alpha_2 = \alpha_1$.

Lemma 1.11.2. _If_ $\tau: L_1 \longrightarrow L_2$ _is a map of line bundles, it is a map of line bundles with connection if and only if for any_ $s \in S_2$ _and any_ $p \in X_1$ _and_ $v \in T_p(X_1)$ _one has_

$$\tau(\nabla^1_v \tau^* s) = \nabla^2_{\rho_* v} s \tag{1.11.3}$$

where ∇^i _is covariant differentiation for_ (L^i, α^i) _and_ $\rho = \check{\tau}$.

Proof. Let $U_2 \subseteq X_2$ be any open set with an element $s \in S^*(U_2, L_2)$. Thus $\tau^* s \in S^*(U_1, L_1)$ where $U_1 = \rho^{-1} U_2$ and if $p \in U_1$ and $v \in T_p(X_1)$ one has $\nabla^1_v \tau^* s = 2\pi i \langle \alpha(\tau^* s), v \rangle (\tau^* s)(p)$ and hence $\tau(\nabla^1_v \tau^* s) = 2\pi i \langle \alpha(\tau^* s), v \rangle s(\rho(p))$. On the other hand $\nabla^2_{\rho_* v} s = 2\pi i \langle \alpha(s), \rho_* v \rangle s(\rho(p))$. Thus since clearly

$$\tau^* \phi s = (\phi \circ \rho) \tau^* s$$

for any $\phi \in C(U_2)$, it follows that (1.11.3) holds at all points in U_1 if and only if

$$\rho^*(\alpha(s)) = \alpha(\tau^* s) . \tag{1.11.4}$$

But now if $r = \tau^* s$ then $\sigma_s^{-1} \circ \tau \circ \sigma_r = 1 \times \rho : \underline{\underline{C}}^* \times U_1 \longrightarrow \underline{\underline{C}}^* \times U_2$ so that $(1 \times \rho)^* = \sigma_r^* \tau^* (\sigma_s^{-1})^*$. But then by (1.5.2) and (1.5.5) one has $(1 \times \rho)^* \left(\frac{1}{2\pi i} \frac{dz}{z}, \alpha(s)\right) = \sigma_r^* (\tau^* \alpha_2)$. Thus $\sigma_r^* (\tau^* \alpha_2) = \left(\frac{1}{2\pi i} \frac{dz}{z}, \rho^* \alpha(s)\right)$. On the other hand by (1.5.2) and (1.5.5) one has $\sigma_r^* (\alpha_1) = \left(\frac{1}{2\pi i} \frac{dz}{z}, \alpha(\tau^* s)\right)$. But then $\sigma_r^* \alpha_1 = \sigma_r^* \tau^* \alpha_2$ if and only if (1.11.4) holds. But since σ_r is a diffeomorphism one has that $\tau^* \alpha_2 = \alpha_1$ in $\tilde{\pi}_1^{-1} U_1$ if and only if (1.11.3) holds at all points in U_1 . But since U_1 is arbitrary this proves the lemma. QED

Remark 1.11.3. One has an analogue of the map (1.11.1) for sections along curves. Using the notation above assume $\gamma:[a,b] \longrightarrow X_1$ is a piece-wise smooth curve in X_1 so that $\rho\gamma$ is a piece-wise smooth curve in X_2 . If $\tau: L_1 \longrightarrow L_2$ is a map of line bundles lifting ρ and r is a section along $\rho\gamma$ then $\tau^* r$ is a section along γ where if $t \in [a,b]$ then $(\tau^* r)(t) \in L^1_{\gamma(t)}$ is given by the relation $\tau\big((\tau^* r)(t)\big) = r(t)$. Since clearly

$$\tau^*(fr) = f\tau^* r \tag{1.11.5}$$

for any $f \in C[a,b]$ and since for any $s \in S(X_2, L_2)$ one has

$$\tau^*(s \circ \rho\gamma) = \tau^* s \circ \gamma \tag{1.11.6}$$

it is clear (since locally $r = f(s \circ \rho\gamma)$ where s does not vanish at the point under consideration) that if γ and r are smooth so is $\tau^* r$. We may reformulate Lemma 1.11.2 in terms of parallel transport.

Lemma 1.11.3. <u>With the same assumption as in Lemma</u> 1.11.2, <u>one has that</u> τ <u>is a map of line bundles with connection if and only</u>

if for every smooth curve γ in X_1 and smooth section r along the curve $\rho\gamma$ in X_2 one has

$$\tau^* \nabla r = \nabla \tau^* r \quad . \qquad (1.11.7)$$

Equivalently τ is a map of line bundles with connection if for any piece-wise smooth curve $\gamma:[a,b] \longrightarrow X_1$ one has

$$\tau \circ P^1_\gamma = P^2_{\rho\gamma} \circ \tau \qquad (1.11.8)$$

as an equality of maps $L^1_{\gamma(a)} \longrightarrow L^2_{\gamma(b)}$ where P^1_γ and $P^2_{\rho\gamma}$ are respectively parallel transport along γ and $\rho\gamma$ relative to (L_1, α_1) and (L_2, α_2).

Proof. Assume (1.11.7) holds for all r and γ. Let $p \in X_1$ $v \in T_p(X_1)$ and $s \in S(X_2, L_2)$. Then if γ is a smooth curve in X_1 such that $\gamma(t_o) = p$ and $\gamma'(t_o) = v$ and r is the section $s \circ \rho\gamma$ along $\rho\gamma$ then upon applying τ to (1.11.7) one obtains $\nabla^2_{\rho_* v} s = \tau(\nabla^1_v \tau^* s)$ by (1.7.2) and (1.11.6). But then τ is a map of line bundles with connection by Lemma 1.11.2.

Conversely assume τ is a map of line bundles with connection and let γ be a smooth curve in X_1 and r a nowhere vanishing section along $\rho\gamma$. Then since $\tau((\tau^* r)(t)) = r(t)$ one has $\tau_* ((\tau^* r)'(t)) = r'(t)$. Thus $\langle \tau^* \alpha_2, (\tau^* r)'(t) \rangle = \langle \alpha_2, r'(t) \rangle$. But $\tau^* \alpha_2 = \alpha_1$ by assumption and hence $\langle \alpha_1, (\tau^* r)'(t) \rangle = \langle \alpha_2, r'(t) \rangle$. But $\nabla r = 2\pi i \langle \alpha_2, r' \rangle r$ by (1.7.3) and hence $\tau^* \nabla r = 2\pi i \langle \alpha_1, (\tau^* r)' \rangle \tau^* r$ by (1.11.5) and hence $\tau^* \nabla r = \nabla \tau^* r$ by (1.7.3). This proves (1.11.7) for a nowhere vanishing section. The general case follows immediately from (1.7.4).

Now let $\gamma:[a,b]\longrightarrow X_1$ be a piece-wise smooth curve and let $\gamma_t = \gamma|[a,t]$ for $t \in [a,b]$. Then if $0 \neq x \in (L_1)_{\gamma(a)}$ and $y = \tau x \in (L_2)_{\rho\gamma(a)}$ one defines a non-trivial auto-parallel section r along $\rho\gamma$ by putting $r(t) = P^2_{\rho\gamma_t}\, y$. But now if (1.11.8) holds for all piece-wise curves then one has $r(t) = \tau P^1_{\gamma_t}\, x$. But s is a non-trivial auto-parallel section along γ where $s(t) = P^1_{\gamma_t}\, x$. Thus $\tau s(t) = r(t)$ and hence $s = \tau^* r$ is auto-parallel. Thus if γ is smooth then since both sides vanish, (1.11.7) holds for a non-trivial auto-parallel section r along $\rho\gamma$. But then using (1.7.4), it follows that (1.11.7) holds for all smooth sections along $\rho\gamma$ since any such section is r multiplied by a smooth function on $[a,b]$.

Finally (1.11.7) implies (1.11.8) along smooth curves, since (1.11.7) implies that r is auto-parallel if and only if $\tau^* r$ is auto-parallel and $\tau\big((\tau^* r)(t)\big) = r(t)$. But (1.11.8) for smooth curves obviously implies (1.11.8) for piece-wise smooth curves. QED

1.12.1. Given manifolds X_i, $i = 1,2$, a line bundle L_2 over X_2 and a smooth map $\rho: X_1\longrightarrow X_2$, as is well known, one may define in a natural way a line bundle $\rho^* L_2$ over X_1 and a map $\tau_\rho:\rho^*(L_2)\longrightarrow L_2$ of line bundles which is a lifting of ρ. Indeed one lets $\rho^*(L_2)$ be the submanifold of $L_2 \times X_1$ defined, set-wise, as the union over all $p \in X_1$ of

$$\big(\rho^*(L_2)\big)_p = \big((L_2)_{\rho(p)}, p\big) \subseteq L_2 \times X_1 . \qquad (1.12.1)$$

Clearly $(L_2)_{\rho(p)}$ induces a linear structure on $\rho^*(L_2)$ and if $\{(U_i,s_i)\}$, $i \in I$ is a local system for L_2 then $(\rho^{-1}U_i, r_i)$, i I , is a local system for $\rho^*(L_2)$ where $r_i(p) = \left(s_i\left(\rho(p)\right), p\right)$. One defines τ_ρ by the relation $\tau_\rho(x,p) = x$ for $x \in (L_2)_{\rho(p)}$. It is obvious that τ_ρ is a map of line bundles and $\check{\tau}_\rho = \rho$.

Proposition 1.12.1. <u>Assume</u> (L_2,α_2) <u>is a line bundle with connection over</u> X_2 . <u>Then if</u> L_1 <u>is a line bundle over</u> X_1 <u>and</u> $\tau: L_1 \longrightarrow L_2$ <u>is a lifting of a smooth map</u> $\rho: X_1 \longrightarrow X_2$, <u>then</u> $(L_1, \tau^*\alpha)$ <u>is a line bundle with connection over</u> X_1 <u>and</u> τ <u>is a map of line bundles with connection.</u>

<u>In particular</u> $\left(\rho^*(L_2), \tau_\rho^* \alpha_2\right)$ <u>is a line bundle with connection over</u> X_1 <u>and</u> τ_ρ <u>is a map of line bundles with connection.</u>

<u>Proof</u>. Since τ^* is linear at each point it is obvious that $\tau^*\alpha$ satisfies (1) and (2) of § 1.5. QED

Now if $\rho: X_1 \longrightarrow X_2$ is a given smooth map and (L_i, α_i), $i = 1,2$ are line bundles with connection over X_i, respectively, we will say it is liftable relative to $(L_i\ \alpha_i)$, $i = 1,2$, if there is a map $\tau: L_1 \longrightarrow L_2$ of line bundles with connection such that $\check{\tau} = \rho$. Instead of the cohomological condition given in Remark 1.11.1 one has

Proposition 1.12.2. <u>Let</u> Γ_i, $i = 1,2$ <u>be the set of all piece-wise smooth curves on</u> X_i <u>and let</u> $Q_i: \Gamma_i \longrightarrow \underline{\underline{C}}^*$ <u>be the scalar parallel transport function defined by</u> (L_i, α_i) . (<u>See</u> (1.8.2)).

Then if $\rho: X_1 \longrightarrow X_2$ *is any smooth map it is liftable to a map* $\tau: L_1 \longrightarrow L_2$ *of line bundles with connection if and only if*

$$Q_2(\rho\gamma) = Q_1(\gamma) \tag{1.12.2}$$

for all $\gamma \in \Gamma_1$. *Moreover* τ *is unique up to global multiplication by a constant* $c \in \mathbb{C}^*$ *in the sense that* τ' *is another such lifting if and only if* τ' *is of the form*

$$\tau' = \tau \circ c . \tag{1.12.3}$$

That is, $\tau'(x) = \tau(c\, x)$ *for all* $x \in L_1$.

Furthermore if L_i *possesses an* α_i*-invariant Hermitian structure* H_i, $i = 1,2$ (*i.e. the* Q_i *are* $\mathbb{T}$*-valued by Proposition* 1.9.2) *then* τ *can be chosen so that on* L_1^*, $|H_2|^2 \circ \tau = |H_1|^2$. (*That is,* $\tau|(L_1)_p$ *is an isometry for all* $p \in X_1$.) *In such a case* τ *is unique up to global multiplication of* L_1 *by a constant in* $\mathbb{T}$.

Proof. If ρ is liftable to a map $\tau: L_1 \longrightarrow L_2$ of line bundles with connection, then by (1.11.8) one has $\tau \circ P^1_\gamma = P^2_{\rho\gamma} \circ \tau$ for all piece-wise smooth curves γ in X_1. But $\gamma \in \Gamma_1$ where $\gamma: [a,b] \longrightarrow X_1$ and if $0 \neq x \in (L_1)_{\gamma(a)}$ then one has $P^1_\gamma x = Q_1(\gamma)\, x$. But $\rho\gamma$ is also closed in X_2 and $P^2_{\rho\gamma}\, y = Q_2(\rho\gamma)\, y$ where $y = \tau x$. Thus $\tau P^1_\gamma x = Q_1(\gamma)\, y = P^2_{\rho\gamma}\, y = Q_2(\rho\gamma)\, y$. Hence $Q_1(\gamma) = Q_2(\rho\gamma)$.

Now conversely assume (1.12.2) is satisfied. Now if $L = \rho^*(L_2)$ and $\alpha = \tau_\rho^* \alpha_2$ then (L,α) is a line bundle with connection over X_1 and τ_ρ is a map of line bundles with connection. Thus from above one has $Q(\gamma) = Q_2(\rho\gamma)$ for $\gamma \in \Gamma_1$, where Q is the scalar parallel transport function relative to (L,α). But then $Q_1 = Q$. Now to show that there exists a lifting τ of ρ relative to (L_i,α_i), $i = 1,2$, it suffices to show that there exists a lifting $\sigma: L_1 \longrightarrow L$ of the identity map of X_1 relative to (L_1,α_1) and (L,α) since in such a case we put $\tau = \tau_\rho \circ \sigma$.

We define σ by first defining σ_o for a fixed point $o \in X_1$ to be an arbitrary linear isomorphism $\sigma_o: (L_1)_o \longrightarrow L_o$. Now let $p \in X_1$ be arbitrary and let γ be any piece-wise smooth curve joining o and p. Also let P_γ and P^1_γ be parallel transport along γ relative to (L,α) and (L_1,α_1). We define $\sigma_p:(L_1)_p \longrightarrow L_p$ by the relation $\sigma_p = P_\gamma \circ \sigma_o \circ (P^1_\gamma)^{-1}$. The relation $Q(\gamma_1) = Q_1(\gamma_1)$ for all closed piece-wise curves γ_1 on X_1 then implies that σ_p is independent of γ and hence a map $\sigma: L_1 \longrightarrow L$ is defined so that $\sigma|(L_1)_p = \sigma_p$. To show that σ is a lifting we have only to show that σ is smooth. For this it suffices to show that for any $p \in X$ there exists a neighborhood U of p and an element $w \in S^*(U,L_1|U)$ such that the section σw of L, over U, is smooth where $(\sigma w)(q) = \sigma\big(w(q)\big)$. But if we choose w to be the section given by Lemma 1.9.1 then by definition of σ clearly σw also satisfies the condition of Lemma 1.9.1 and hence, by that lemma, σw is smooth. Thus σ is a map of line bundles. But it is also a map of line bundles with connection since by definition it clearly satisfies (1.11.8).

Thus a map $\tau: L_1 \longrightarrow L_2$ of line bundles with connection exists such that $\tau = \rho$. Now assume τ' is another such map. Then by Lemma 1.11.1 there exists a line bundle equivalence $\mu: L_1 \longrightarrow L_1$ of L_1 with itself such that $\tau \circ \mu = \tau'$. But then $\mu^* \tau^* \alpha_2 = \tau'^* \alpha_2$. Hence $\mu^* \alpha_1 = \alpha_1$ so that μ is an equivalence of line bundles with connection. But by Corollary 1 to Lemma 1.10.1 the map μ is just multiplication by a constant in $\underline{\underline{C}}^*$, proving (1.12.3).

Finally assume H_i, $i = 1,2$ are, respectively, α_i-invariant Hermitian structures in L_i and $\tau: L_1 \longrightarrow L_2$ is a lifting of ρ relative to (L_i, α_i), $i = 1,2$. Then since $\tau^* \alpha_2 = \alpha_1$ one has on L_1^* (see (1.9.2))

$$\tau^* \left(\frac{d|H_2|^2}{|H_2|^2} \right) = \frac{d|H_1|^2}{|H_1|^2} .$$

But then one has $d\left(\frac{|H_2|^2 \circ \tau}{|H_1|^2} \right) = 0$, or that $|H_2|^2 \circ \tau = c|H_1|^2$ where c is a positive number. Hence if τ is replaced by $\tau \circ c^{-\frac{1}{2}}$ then $|H_2|^2 \circ \tau = |H_1|^2$ on L_1^* so that τ_p is an isometry for all p. As such τ is now clearly unique up to scalar in $\underline{\underline{T}}$. QED

Proposition 1.12.3. <u>Let</u> (L_i, α_i), $i = 1,2$, <u>be line bundles with connection over a manifold</u> X <u>and let</u> $Q_i: \Gamma \longrightarrow \underline{\underline{C}}^*$ <u>be the corresponding scalar parallel transport function. Then</u> (L_1, α_1) <u>and</u> (L_2, α_2) <u>are equivalent if and only if</u> $Q_1 = Q_2$.

<u>Proof</u>. We simply apply Proposition 1.12.2 to the case where $\rho: X \longrightarrow X$ is the identity map. QED

It follows from Proposition 1.12.3 that the parallel transport function corresponding to (L,α) depends only on the equivalence class $\ell = [(L,\alpha)]$ of (L,α) thereby defining $Q^{\ell}\colon \Gamma \longrightarrow \underline{\underline{C}}^{*}$ for all equivalence classes ℓ.

Given a manifold X we let $\mathcal{L}_c = \mathcal{L}_c(X)$ be the set of all equivalence classes $\ell = [(L,\alpha)]$ such that Q^{ℓ} is $\underline{\underline{T}}$-valued (that is by Proposition 1.9.2. there exists an α-invariant Hermitian structure in L).

Remark 1.12.1. One notes that by Proposition 1.12.3 a class $\ell \in \mathcal{L}_c$ not only determines Q^{ℓ} but it is also completely determined by Q^{ℓ}.

Now note that if X and Y are manifolds and $\rho\colon X \longrightarrow Y$ is a smooth map then the correspondence $(L,\alpha) \longrightarrow (\rho^{*}L, \tau_{\rho}^{*}\alpha)$ defines a map

$$\mathcal{L}_c(Y) \longrightarrow \mathcal{L}_c(X) \tag{1.12.4}$$

$\ell \longrightarrow \rho^{*}\ell$ where if $\ell = [(L,\alpha)]$ then $\rho^{*}\ell = [(\rho^{*}L, \tau_{\rho}^{*}\alpha)]$. That it depends only on the equivalence class is clear for example from Propositions 1.12.2 and 1.12.3. They also imply that

$$Q^{\rho^{*}\ell}(\gamma) = Q^{\ell}(\rho\gamma) \tag{1.12.5}$$

for any piece-wise smooth curve γ on X.

Remark 1.12.2. Note that if $[(L_i,\alpha_i)] = \ell_i \in \mathcal{L}_c(X_i)$, $i = 1,2$, and $\tau\colon L_1 \longrightarrow L_2$ is a map of line bundles with connection, by Proposition 1.12.2, one has $(\check{\tau})^{*}\ell_2 = \ell_1$.

1.13. Now for any manifold X we let $D(X)$ be the group of all diffeomorphisms of X. Then $D(X)$ operates on $\mathcal{L}_c(X)$ by defining $g \cdot \ell = (g^{-1})^* \ell$ for any $g \in D(X)$, $\ell \in \mathcal{L}_c(X)$ so that by (1.12.5) one has for any $\gamma \in \Gamma(X)$

$$Q^{g \cdot \ell}(\gamma) = Q^{\ell}(g^{-1}\gamma) \quad . \tag{1.13.1}$$

Now for any $\ell \in \mathcal{L}_c$ let $D_\ell = D_\ell(X) \subseteq D(X)$ be the isotropy group at ℓ. That is D_ℓ is the group of all diffeomorphisms g of X such that for all closed piece-wise smooth curves γ on X one has

$$Q^{\ell}(\gamma) = Q^{\ell}(g^{-1}\gamma) \quad . \tag{1.12.1}$$

On the other hand if (L,α) is any line bundle with connection over X where L has an α-invariant Hermitian structure H, let $E(L,\alpha)$ be the group of all diffeomorphisms $\tau \colon L \longrightarrow L$ which (1) are maps of line bundles with connection (and hence define an equivalence of (L,α) with itself) and (2) satisfy $|H|^2 \circ \tau = |H|^2$.

Thus (see § 1.11) if $\tau \in E(L,\alpha)$ it defines $\check{\tau} \in D(X)$ and τ induces an isometry

$$\tau_p : L_p \longrightarrow L_{\check{\tau}(p)} \quad .$$

Remark 1.13.1. By Proposition 1.9.1 note that $E(L,\alpha)$ is independent of the choice of H.

Theorem 1.13.1. <u>Let</u> (L,α) <u>be a line bundle with connection over a manifold</u> X <u>where</u> L <u>admits an</u> α-<u>invariant Hermitian structure</u>.

Let ℓ *be the equivalence class* $[(L,\alpha)]$ *so that* $\ell \in \mathcal{L}_c$. *(See* § 1.12). *Then one has an exact sequence of groups*

$$1 \longrightarrow \underline{\underline{T}} \longrightarrow E(L,\alpha) \longrightarrow D_\ell(X) \longrightarrow 1$$

making $E(L,\alpha)$ *a central extension of* $D_\ell(X)$ *by the circle group* $\underline{\underline{T}}$. *Here the injection* $\underline{\underline{T}} \longrightarrow E(L,\alpha)$ *is defined by the scalar action of* $\underline{\underline{T}}$ *on* L . *The map* $E(L,\alpha) \longrightarrow D_\ell(X)$ *is given by* $\tau \longrightarrow \check{\tau}$ (*see* § 1.11).

Proof. If $\tau \in E(L,\alpha)$ then $\check{\tau} \in D_\ell(X)$ by Remark 1.12.2. But this map is also surjective by Proposition 1.12.2 (the lifting is clearly a diffeomorphism). Furthermore Proposition 1.12.2 asserts that $\check{\tau}_1 = \check{\tau}_2$ where $\tau_i \in E(L,\alpha)$, $i = 1,2$, if and only if the τ_i differ by the action of a scalar in $\underline{\underline{T}}$. QED

If $g \in D_\ell(X)$ then by a lifting of g to $E(L,\alpha)$ we mean an element $\tau \in E(L,\alpha)$ such that $\check{\tau} = g$.

Now assume a group G operates on a group of diffeomorphisms of X and an element $\ell \in \mathcal{L}_c(X)$ is fixed under G . That is one has a homomorphism

$$\sigma\colon G \longrightarrow D_\ell(X) .$$

Then by lifting of σ to $E(L,\alpha)$ we mean a homomorphism $\nu\colon G \rightarrow E(L,\alpha)$ such that the diagram

$$\begin{array}{ccccccccc} 1 & \longrightarrow & \underline{\underline{T}} & \longrightarrow & E(L,\alpha) & \longrightarrow & D_\ell(X) & \longrightarrow & 1 \\ & & & & & \nwarrow & \uparrow \sigma & & \\ & & & & & & G & & \end{array} \tag{1.13.1}$$

is commutative.

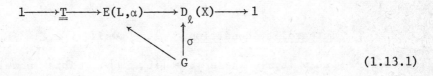

Remark 1.13.1. The homomorphism σ defines a cohomology class $[\mu] \in H^2(G, \underline{\underline{T}})$ (trivial action) where μ is a cocycle given by $\mu(a,b) = \mu_o(a)\, \mu_o(b)\, \mu_o(ab)^{-1}$ $\underline{\underline{T}}$ where $\mu_o: G \longrightarrow E(L,\alpha)$ is any map making (1.13.1) commutative. A lifting ν of σ exists if and only if $[\mu] = 0$. That is, if and only if there exists a map $\rho: G \longrightarrow \underline{\underline{T}}$ such that $\mu(a,b) = \rho(ab)\left(\rho(b)\, \rho(a)\right)^{-1}$.

If a group H operates on a set K then K is called a principal homogeneous space for H if given $k, \ell \in K$ there exists a unique $h \in H$ such that $h \cdot k = \ell$.

Proposition 1.13.1. _If_ $\nu: G \longrightarrow E(L,\alpha)$ _is any lifting of_ σ _then the most general lifting of_ σ _is uniquely of the form_ $a \longrightarrow \chi(a)\, \nu(a)$ _where_ χ _is a character of_ G. _That is_ χ _is a homomorphism_

$$\chi : G \longrightarrow \underline{\underline{T}} .$$

Thus the set of liftings is a principal homogeneous space for the character group G^* _of_ G _where_ $\chi\nu$ _is the lifting given by_ $(\chi\nu)(a) = \chi(a)\, \nu(a)$.

Proof. Since $\underline{\underline{T}}$ is central in $E(L,\alpha)$ it is obvious from Theorem 1.13.1 that any two liftings differ by a character. QED

2. _The integrality condition._ 1. Now if $\ell \in \mathcal{L}_c(X)$, we define $\operatorname{curv} \ell = \operatorname{curv}(L,\alpha)$ where $\ell = [(L,\alpha)]$. This is well-defined since, clearly by Proposition 1.6.1 one has $\operatorname{curv}[(L_1,\alpha_1)] = \operatorname{curv}[(L_2,\alpha_2)]$ if $[(L_1,\alpha_1)] = [(L_2,\alpha_2)]$. We

recall that by Corollary 1 to Proposition 1.9.1, curv ℓ, for any $\ell \in \mathcal{L}_c(X)$ is a closed real 2-form on X .

Now for any closed 2-form ω let $\mathcal{L}_c(X,\omega)$ be the set of all $\ell \in \mathcal{L}_c$ such that curv $\ell = \omega$. Thus

$$\mathcal{L}_c(X) = \bigcup \mathcal{L}_c(X,\omega)$$

is a disjoint union, over all closed real 2-forms on X .

Now recall that the de Rham cohomology group $H^2_D(X, \underline{\underline{R}})$ is defined as the quotient of the space of all smooth, real closed 2-forms on X by the space of all smooth real exact 2-forms. One also knows that $H^2_D(X,\underline{\underline{R}})$ is naturally isomorphic to the Cech cohomology group $H^2(X, \underline{\underline{R}})$ considered in § 1.2. We recall how the isomorphism may be obtained.

Let $\omega \in \Omega^2(X)$ be real and closed, and let $\mathcal{U} = \{U_i\}$, $i \in I$, be any contractible covering of X . Now since U_i is contractible we can write

$$\omega = d\alpha_i \qquad\qquad \text{in } U_i$$

where $\alpha_i \in \Omega^1(U_i)$ is real. On the other hand since $U_i \cap U_j$ is contractible, and since $d(\alpha_j - \alpha_i) = \omega - \omega = 0$ on $U_i \cap U_j$, we can write

$$\alpha_j - \alpha_i = d\, f_{ij}$$

where $f_{ij} \in C(U_i \cap U_j)$ is real. But obviously $df_{ij} + d\, f_{jk} - d\, f_{ik} = d(f_{ij} + f_{jk} - f_{ij}) = 0$ on $U_i \cap U_j \cap U_k$ and since this space is contractible one has

$$f_{ij} + f_{jk} - f_{ik} = a_{ijk}$$

is a real constant if $U_i \cap U_j \cap U_k \neq \emptyset$. This defines a class $[a] \in H^2(X, \underline{\underline{R}})$. It follows easily that the class $[a]$ is independent of the choice of covering, the choice of the α_i or of the f_{ij} and depends only on the class $[\omega] \in H^2(X, \underline{\underline{R}})$ defining a homomorphism

$$H^2_D(X, \underline{\underline{R}}) \longrightarrow H^2(X, \underline{\underline{R}}) . \tag{2.1.1}$$

The partition of unity argument in the proof of the isomorphism (1.2.3) now used, (twice) injectively and surjectively, in a similar way easily shows that the map (2.1.1) is an isomorphism and we shall henceforth identify $H^2_D(X, \underline{\underline{R}})$ with $H^2(X, \underline{\underline{R}})$.

If A and B are abelian groups then any homomorphism $\varepsilon: A \longrightarrow B$ induces a homomorphism $\varepsilon: H^2(X,A) \longrightarrow H^2(X,B)$ where $\varepsilon[a] = [\varepsilon a]$ and $(\varepsilon a)_{ijk} = \varepsilon(a_{ijk})$. In particular the injection $\varepsilon: \underline{\underline{Z}} \longrightarrow R$ induces a homomorphism

$$\varepsilon: H^2(X, \underline{\underline{Z}}) \longrightarrow H^2(X, \underline{\underline{R}}) . \tag{2.1.2}$$

A class $\gamma \in H^2(X, \underline{\underline{R}})$ is called integral in case γ lies in the image of the map.

Remark 2.1.1. One notes from above that if ω is a closed real 2-form, then $[\omega] \in H^2(X, \underline{\underline{R}})$ is integral if and only if there exists a contractible covering $\mathcal{U} = \{U_i\}$, $i \in I$, of X, real 1-forms $\alpha_i \in \Omega^1(U_i)$, and functions $f_{ij} \in C(U_i \cap U_j)$ such that $\omega = d\alpha_i$ in

U_i, $\alpha_j - \alpha_i = d\, f_{ij}$ and such that if $U_i \cap U_j \cap U_k \neq \emptyset$ then $f_{ij} + f_{jk} - f_{ik} = a_{ijk}$ is an integer.

Recall that $\mathcal{L}$ is the set of equivalence classes of line bundles over X and one has a bijection $\rho: \mathcal{L} \longrightarrow H^2(X, \underline{\underline{Z}})$ (see (1.2.3)). On the other hand one clearly has a map

$$\mathcal{L}_c \longrightarrow \mathcal{L} \tag{2.1.3}$$

given by $[(L,\alpha)] \longrightarrow [L]$.

One now has (see also Weil [5], Lemma 2, p. 90),

Proposition 2.1.1. <u>Let</u> ω <u>be any closed real 2-form on</u> X. <u>Then</u> $\mathcal{L}_c(X,\omega)$ <u>is not empty if and only if</u> $[\omega] \in H^2(X, \underline{\underline{R}})$ <u>is integral. Moreover in such a case the image of</u> $\mathcal{L}_c(X,\omega)$ <u>under the map</u> (2.1.3) <u>is the set of all</u> $[L] \in \mathcal{L}$ <u>such that</u>

$$\varepsilon\rho[L] = [\omega] .$$

(<u>That is, identifying</u> $\mathcal{L}$ <u>with</u> $H^2(X, \underline{\underline{Z}})$, <u>the image of</u> $\mathcal{L}_c(X,\omega)$ <u>is the inverse of</u> $[\omega]$ <u>under the map</u> (2.1.2).

<u>Proof</u>. Assume $\mathcal{L}_c(X,\omega)$ is not empty and $[(L,\alpha)] \in \mathcal{L}_c(X,\omega)$. Let H be an α-invariant Hermitian structure in L and let $\{(U_i, s_i)\}$, $i \in I$, be a local system for L where the covering is contractible. By replacing s_i with $\frac{s_i}{|s_i|}$ we may assume $|s_i| = 1$. But then one has $|c_{ij}| = 1$ for the transition functions c_{ij} (see § 1.1.). Furthermore since $\xi|s_i|^2 = 0$ for all $\xi \in \underline{\mathcal{U}}$ it follows from (1.9.3)

that $\alpha_i = \alpha(s_i)$ is real for all i. On the other hand $f_{ij} = \frac{1}{2\pi i} \log c_{ij}$ is real and since in $U_i \cap U_j \cap U_k$, $c_{ij}\, c_{jk} = c_{ik}$ it follows that $f_{ij} + f_{jk} - f_{ik} = a_{ijk}$ is an integer if $U_i \cap U_j \cap U_k \neq \emptyset$. But now since $\alpha_j - \alpha_i = \frac{1}{2\pi i} \frac{dc_{ij}}{c_{ij}}$ by (1.4.5) one has $\alpha_j - \alpha_i = d\, f_{ij}$. However $\omega = d\alpha_i$ in U_i by (1.6.2). Thus $[\omega]$ is integral by Remark 2.1.1. Also $\rho[L] = [a]$ by § 1.2 where a is the integral Cech co-cycle given by $ijk \longrightarrow a_{ijk}$. On the other hand by (2.1.2) one has $[\omega] = \varepsilon[a]$. Thus $\varepsilon\, \rho[L] = [\omega]$.

Now conversely assume $[\omega]$ is integral. Let U_i, $i \in I$, α_i, f_{ij} and a_{ijk} be as in Remak 2.1.1. But since $f_{ij} + f_{jk} - f_{ik} = a_{ijk}$ is an integer, one has $c_{ij}\, c_{jk} = c_{ik}$ in $U_i \cap U_j \cap U_k$ and $c_{ij} = c_{ij}^{-1}$ where one defines $c_{ij} = e^{2\pi i\, f_{ij}}$. But then there exists a line bundle L with local system $\{(U_i, s_i)\}$, $i \in I$, having the c_{ij} as transition functions. Clearly $\varepsilon(\rho[L]) = [\omega]$. On the other hand if L_1 is any line bundle over X such that $\varepsilon\rho[L_1] = [\omega]$ we could choose the f_{ij} so that $[L_1] = [L]$. Indeed if the c_{ij}^1 are transition functions of L_1 relative to the covering $\{U_i\}$ (see Remark 1.2.1) and $f_{ij}^1 = \frac{1}{2\pi i} \log c_{ij}^1$, then the assumption $\varepsilon\rho[L_1] = [\omega]$ implies that $\varepsilon[a^1] = \varepsilon[a]$ where $a_{ijk}^1 = f_{ij}^1 + f_{jk}^1 - f_{ik}^1$. But then replacing f_{ij} by $f_{ij} + r_{ij}$ for a suitable real Cech cochain r, we have $a = a^1$ so that $[a] = [a^1]$ and hence $[L] = [L_1]$ since $\rho[L] = [a]$ and $\rho[L_1] = [a_1]$.

Now the relation $\alpha_j - \alpha_i = d\, f_{ij}$ implies that

$\alpha_j - \alpha_i = \frac{1}{2\pi i} \frac{d\, c_{ij}}{c_{ij}}$ in $U_i \cap U_j$ and hence by Corollary 1 to Proposition 1.4.1 there exists a connection form α in L^* such that $\alpha(s_i) = \alpha_i$. But since $d\alpha_i = \omega$ in U_i one has $\omega = \text{curv}\, \ell$ where $\ell = [(L,\alpha)]$. We have only to show that $\ell \in \mathcal{L}_c$ or that (L,α) has an α-invariant Hermitian structure H. Let $x,y \in L$ where $p = \pi(x) = \pi(y) \in X$. Then $p \in U_i$ for some $i \in I$. Put $H(x,y) = \frac{x}{s_i(p)} \left(\overline{\frac{y}{s_i(p)}}\right)$ (see § 1.4 for notation). This does not depend upon $i \in I$ since $|c_{ij}| = 1$. The smoothness of $|H|^2$ is obvious so that H defines a Hermitian structure in L. One has $|s_i|^2 = 1$ in U_i so that to show that H is α-invariant it clearly suffices using (1.4.2) to show that for all real $\xi \in \mathfrak{M}$, $(\nabla_\xi s_i, s_i) + (s_i, \nabla_\xi s_i) = 0$. But this is immediate from (1.4.3) since α_i is real. QED

2.2. Now let the pair (X,ω) denote a manifold X together with a real closed 2-form ω. By Proposition 2.1.1 we know that $\mathcal{L}_c(X,\omega)$ is not empty if and only if $[\omega] \in H^2(X, \mathbb{R})$ is integral. In such a case we wish to give more information about $\mathcal{L}_c(X,\omega)$. In the simply connected case one has

Theorem 2.2.1. _Let_ (X,ω) _be given where_ $[\omega]$ _is integral. Assume_ X _is simply connected. Then_ $\mathcal{L}_c(X,\omega)$ _has exactly one element. In fact, up to equivalence there exists a unique line bundle with_

<u>connection</u> (L,α) . <u>Moreover there exists an</u> α-<u>invariant Hermitian structure in</u> L . <u>For the unique element</u> $\ell = [(L,\alpha)] \in \mathcal{L}_c(X,\omega)$ <u>one has for any closed piece-wise smooth curve</u> γ <u>in</u> X

$$Q^{\ell}(\gamma) = e^{-2\pi i \int_{\sigma} \omega} \qquad (2.2.1)$$

<u>where</u> σ <u>is any surface of deformation of</u> γ (<u>see</u> § 1.8).

<u>Proof</u>. By Proposition 2.1.1 $\mathcal{L}_c(X,\omega)$ is not empty. But if $\ell \in \mathcal{L}_c(X,\omega)$ then by Remark 1.12.1 ℓ is determined by Q^{ℓ} . But by Theorem 1.8.1, $Q^{\ell}(\gamma)$ is given explicity, independent of ℓ , for those γ which are homotopic to a point. However since X is simply connected this is true of all γ . Hence ℓ is unique. QED

Now let $D(X,\omega)$, the generalized "symplectic" group, be the subgroup of all $g \in D(X)$ such that $g^{*}\omega = \omega$. We recall that if $\ell \in \mathcal{L}_c(X)$ then $D_{\ell}(X)$ is the group of all $s \in D(X)$ such that $g \cdot \ell = \ell$.

Proposition 2.2.1. <u>Let</u> $\ell \in \mathcal{L}_c(X,\omega)$. <u>Then</u>

$$D_{\ell}(X) \subseteq D_{\ell}(X,\omega)$$

<u>and if</u> X <u>is simply connected one has equality</u>, $D_{\ell}(X) = D(X,\omega)$.

<u>Proof</u>. Let $[(L,\alpha)] = \ell$ and $g \in D_{\ell}(X)$. Then if $\tau \in E(L,\alpha)$ is a lifting of g one has $\tau^{*}\alpha = \alpha$ so that $\tau^{*}d\alpha = d\alpha$. Thus $g^{*}\omega = \omega$. But if $g \in D(X,\omega)$ and X is simply connected, then $Q^{\ell}(g^{-1}\gamma) = Q^{\ell}(\gamma)$ for any $\gamma \in \Gamma$ since $Q^{\ell}(\gamma)$ is given by (2.2.1). Thus $g \in D_{\ell}(\omega)$ by (1.12.1). QED

Theorem 2.2.2. <u>Assume</u> X <u>is simply connected and assume</u> $[\omega]$ <u>is integral so that a line bundle with connection</u> (L,α) <u>exists (uniquely up to equivalence) such that</u> <u>curv</u> $(L,\alpha) = \omega$. <u>Then the map</u> $\tau \longrightarrow \check{\tau}$ <u>defines</u> $E(L,\alpha)$ <u>as a central extension</u>

$$1 \longrightarrow \underline{\underline{T}} \longrightarrow E(L,\alpha) \longrightarrow D(X,\omega) \longrightarrow 1$$

<u>of</u> $D(X,\omega)$ <u>by the circle group</u> $\underline{\underline{T}}$.

<u>Proof</u>. This follows from Proposition 2.2.1 and Theorem 1.13.1.

QED

2.3. Now for the general case let Π be the fundamental group and Π^* the group of all homomorphisms (characters)

$$\chi : \Pi \longrightarrow \underline{\underline{T}}$$

Since any element $\gamma \in \Gamma$ uniquely defines a conjugacy class in Π (without specifying a base point) and since any $\chi \in \Pi^*$ is constant on conjugacy classes we can regard Π^* as the set of all maps $\chi : \Gamma \longrightarrow \underline{\underline{T}}$ such that $\chi(\gamma) = 1$ if γ is homotopic to a point and such that

$$\chi(\gamma_1 + \gamma_2) = \chi(\gamma_1)\ \chi(\gamma_2) \tag{2.3.1}$$

where $\gamma_1 + \gamma_2$ is the usual addition of curves with the same initial and end points.

Assume $[\omega]$ is integral and now let $\mathcal{P}$ be the set of maps $Q : \Gamma \longrightarrow \underline{\underline{T}}$ satisfying (2.3.1) and such that $Q(\gamma)$ is given by (1.8.3) if γ is homotopic to a point.

By Proposition 2.1.1 $\mathcal{P}$ is not empty. Now clearly the group Π^* operates on $\mathcal{P}$ where if $\chi \in \Pi^*$, $Q \in \mathcal{P}$ one puts

$$(\chi \cdot Q)(\gamma) = \chi(\gamma)\, Q(\gamma) . \tag{2.3.2}$$

Proposition 2.3.1. $\mathcal{P}$ <u>is a principal homogeneous space for</u> Π^*. <u>That is, given any</u> $Q_1, Q_2 \in \mathcal{P}$ <u>there exists a unique</u> $\chi \in \Pi^*$ <u>such that</u> $\chi \cdot Q_1 = Q_2$.

<u>Proof</u>. If $Q_1, Q_2 \in \mathcal{P}$ and $\chi : \Gamma \longrightarrow \underline{\underline{T}}$ is defined by $\chi(\gamma) = Q_2(\gamma)/Q_1(\gamma)$ then clearly (2.3.1) is satisfied and $\chi(\gamma) = 1$ if γ is homotopic to a point. Thus $\chi \in \Pi^*$. QED

2.4. Now let X^1 be the simply-connected covering space of X and let

$$\beta : X^1 \longrightarrow X$$

be the covering map. Of course X^1 is a manifold and β is a local diffeomorphism. Also the fundamental group Π of X operates as deck diffeomorphisms of X^1. Writing this action on the right one thus has a map

$$X^1 \times \Pi \longrightarrow X^1 \qquad (q,b) \longrightarrow q \cdot b$$

for $q \in X^1$, $b \in \Pi$.

Now let $\omega^1 = \beta^* \omega$. Obviously ω^1 is invariant under the action of Π and hence this action induces a homomorphism

$$\sigma : \Pi \longrightarrow D(X^1, \omega^1) \tag{2.4.1}$$

where $\sigma(b)\, q = q \cdot b^{-1}$.

Now if $[\omega]$ is integral then clearly $[\omega^1]$ is integral. (The converse is not necessarily true). Assume, in this section, only that $[\omega^1]$ is integral. Let $[(L^1,\alpha^1)] = \ell^1 \in \mathcal{L}_c(X^1,\omega^1)$ be the unique element (Theorem 2.2.1).

Now if a lifting $\nu: \Pi \longrightarrow E(L^1,\alpha^1)$ to σ exists (see § 1.13) we now observe that the space of orbits $L = L^1/\nu(\Pi)$ of Π on L^1 has a structure of a line bundle over X . Indeed first of all the map ν defines a map

$$L^1 \times \Pi \longrightarrow L^1 \qquad\qquad x \longrightarrow x \cdot b$$

for $x \in L^1$, $b \in \Pi$ where $\nu \cdot b = \nu(b)^{-1}\, x$.

Let $\tilde{\beta}: L^1 \longrightarrow L$ be the map given by $x \longrightarrow x \cdot \Pi$. If $U \subseteq X^1$ is a connected open set such that $(U \cdot b) \cap U \neq \emptyset$, for $b \in \Pi$ implies b is the identity, the same is true of $(\pi^1)^{-1}(U) \subseteq L^1$ where $\pi^1: L^1 \longrightarrow X^1$ is the line bundle projection. Thus L has the structure of a manifold, $\tilde{\beta}$ is a covering map and local diffeomorphism. Moreover there clearly exists a unique (smooth) map $\pi: L \longrightarrow X$ such that the diagram

$$\begin{array}{ccc} L^1 & \xrightarrow{\ \pi^1\ } & X^1 \\ {\scriptstyle\tilde{\beta}}\downarrow & & \downarrow{\scriptstyle\beta} \\ L & \xrightarrow{\ \pi\ } & X \end{array} \tag{2.4.2}$$

is commutative. Also if $p \in X$ then there is a unique linear structure in $\pi^{-1}(p)$ such that the map $\tilde{\beta}: (\pi^1)^{-1}(p^1) \longrightarrow \pi^{-1}(p)$, for any $p^1 \in \beta^{-1}(p)$, is a linear isomorphism. The local product structure in $(\pi^1)^{-1}(U)$ for $U \subseteq X^1$, as above and sufficiently small, induces such a structure in $\tilde{\beta}\left((\pi^1)^{-1}(U)\right)$ so that L is a line bundle over X.

Now since α^1 is invariant under Π there exists a unique 1-form α in L^* such that $\tilde{\beta}^*(\alpha) = \alpha^1$ defining the structure of a line bundle with connection (L,α) over X. Clearly $d\alpha = \tilde{\pi}^*\omega$ from (2.4.2) and since Π leaves fixed an α^1-invariant Hermitian structure in L^1 it follows that $[(L,\alpha)] = \ell \in \mathcal{L}_c(X,\omega)$. In particular $[\omega]$ is integral. Since $[(L,\alpha)] \in \mathcal{L}_c(X,\omega)$ depends upon ν we write

$$\ell_\nu = [(L,\alpha)] \ . \tag{2.4.3}$$

Conversely if $[\omega]$ is integral and $[(L,\alpha)] = \ell \in \mathcal{L}_c(X,\omega)$ let L^2 be the line bundle $\tilde{\beta}^*(L)$ over X^1 (see § 1.11) and let $\beta^2: L^2 \longrightarrow L$ be the map $\tilde{\beta}$. The diagram (2.4.2) is thus commutative where β^2 and L^2 replace $\tilde{\beta}$ and L^1. Now if $\alpha^2 = (\beta^2)^*\alpha$ then one has $[(L^2,\alpha^2)] \in \mathcal{L}_c(X^1,\omega^1)$. But by Theorem 2.2.1, (L^2,α^2) and (L^1,α^1) are equivalent and hence $\tilde{\beta}: L^1 \longrightarrow L$ exists so that (2.4.2) is commutative and $\beta^*(\alpha) = \alpha^1$. We define an action of Π on L^1 by putting, for $x \in L^1$, $b \in \Pi$, $x \cdot b$ equal to the unique element $y \in (L^1)_{\pi^1(x) \cdot b}$ such that $\tilde{\beta}(x) = \tilde{\beta}(y)$. Since $\tilde{\beta}(x \cdot b) = \tilde{\beta}(x)$, it follows that α^1 is invariant under Π and since L has an α-invariant Hermitian structure, Π preserves an α^1-invariant Hermitian structure in L^1. Thus the action of Π on L^1 defines a lifting $\nu:\Pi \longrightarrow E(L^1,\alpha^1)$ of σ. This proves

Proposition 2.4.1. <u>If $[\omega^1] \in H^2(X^1, \underline{\underline{R}})$ is integral where $\beta: X^1 \longrightarrow X$ is the covering map for the simply connected covering space of X and $\omega^1 = \beta^*\omega$, then $[\omega] \in H^2(X, \underline{\underline{R}})$ is integral if and only if there is a lifting $\Pi \longrightarrow E(L^1,\alpha^1)$ of the homomorphism $\Pi \longrightarrow D(X^1,\omega^1)$ defined by the deck transformations. Here Π is the fundamental group of X regarded as operating on X^1 and (L^1,α^1) is the unique, up to equivalence, line bundle with connection over X^1 such that $\operatorname{curv}(L^1,\alpha^1) = \omega^1$. That is, in the notation of</u> Remark 1.13.1, $[\omega]$ <u>is integral if and only if the class $[\mu] \in H^2(\Pi, \underline{T})$ vanishes.</u>

2.5. Now assume $[\omega]$ and hence $[\omega^1]$ is integral. Let (L^1,α^1) be as above. Let $p \in X$ and $\gamma \in \Gamma$ be such that p is the initial and end point of γ. Let $p^1 \in \beta^{-1}(p)$ and let γ^1 be the unique piece-wise smooth curve in X^1 which covers γ and has p^1 as the initial point. If the $b \in \Pi$ is the class of γ at p^1 then the end point of γ^1 is $p^1 \cdot b$ and hence (parallel transport)

$$P_{\gamma^1}: (L^1)_{p^1} \longrightarrow (L^1)_{p^1\cdot b} \tag{2.5.1}$$

is a linear isomorphism.

Now let $\nu: \Pi \longrightarrow E(L^2,\alpha^2)$ be any lifting of (2.4.1) and let $[(L,\alpha)] = \ell_\nu \in \mathcal{L}_c(X,\omega)$ be as (2.4.3) so that $L = L^1/\nu(\Pi)$ and $\tilde{\beta}^*\alpha = \alpha^1$. One now has another map

$$\nu(b^{-1}) : (L^1)_{p^1} \longrightarrow (L^1)_{p^1 \cdot b} \quad . \tag{2.5.2}$$

Comparing (2.5.1) and (2.5.2) one has the following expression for scalar parallel transport function Q^{ℓ_ν} on $\Gamma = \Gamma(X)$.

Proposition 2.5. <u>One has for any</u> $0 \neq x \in (L^1)_{p^1}$

$$Q^{\ell_\nu}(\gamma) = \frac{P_{\gamma^1} x}{\nu(b^{-1}) x} \quad . \tag{2.5.3}$$

<u>Proof</u>. Let $y = \tilde{\beta}x \in L_p$. Then since $\tilde{\beta}$ clearly preserves parallel transport one has

$$Q^{\ell_\nu}(\gamma)y = \tilde{\beta}(P_{\gamma^1} x) \ .$$

On the other hand $y = \tilde{\beta}(x \cdot b)$ or $\tilde{\beta}\big(Q^{\ell_\nu}(\gamma)(x{\cdot}b)\big) = Q^{\ell_\nu}(\gamma)\tilde{\beta}(x{\cdot}b)$

$$= Q^{\ell_\nu}(\gamma)y = \tilde{\beta}(P_{\gamma^1} x) \ .$$

Thus $Q^{\ell_\nu}(\gamma)\, x \cdot b = P_{\gamma^1} x$ proving the Proposition since $x \cdot b = \nu(b)^{-1} x$. QED

We recall that by Proposition 2.3.1 the set of lifting of (2.4.1) is a principal homogeneous for Π^* . We can now give the structure of $\mathcal{L}_c(X,\omega)$.

Theorem 2.5.1. <u>Let</u> (X,ω) <u>be a manifold</u> X <u>together with a real closed</u> 2-<u>form</u> ω . <u>Assume</u> $[\omega] \in H^2(X, \underline{R})$ <u>is integral (so that</u> $\mathcal{L}_c(X,\omega)$, <u>the set of equivalence of line bundles with connection having</u> ω <u>as curvature is not empty</u> - <u>see</u> Proposition 2.1.1). <u>Let</u> $\mathcal{P}$ <u>be as in</u> § 2.3 <u>so that</u> $\mathcal{P}$ <u>is a principal homogeneous space for the character group</u> Π^* <u>of the fundamental group</u> Π <u>of</u> λ . <u>Then the map</u>

$$\mathcal{L}_c(X,\omega) \longrightarrow \mathcal{P} \qquad (2.5.4)$$

given by $\ell \longrightarrow Q^{\ell}$ is a bijection making $\mathcal{L}_c(X,\omega)$ a principal Π^* homogeneous space where

$$Q^{\chi\cdot\ell} = \chi Q^{\ell} . \qquad (2.5.5)$$

Furthermore we recall (see Proposition 1.13.1) that the set of liftings ν of (2.4.1) is a principal Π^*-homogeneous space. The correspondence between this set and $\mathcal{L}_c(X,\omega)$ given by $\nu \longrightarrow \ell_\nu$ (see (2.4.3)) is a bijection and is such that

$$\ell_{\chi\nu} = \chi\ell_\nu \qquad (2.5.6)$$

Proof. By Remark 1.12.1 the map (2.5.4) is injective. Now let $\nu:\Pi \longrightarrow E(L^2,\alpha^2)$ be any lifting of σ (see (2.4.1)) so that $Q^{\ell_{\chi\nu}}$ is in the image of (2.5.4) for any $\chi \in \Pi^*$. But by (2.5.3) one has, since $(\chi\nu)(b^{-1}) = \chi(b^{-1})\ \nu(b^{-1})$ for any $\gamma \in \Gamma$

$$\begin{aligned} Q^{\ell_{\chi\nu}}(\gamma) &= \chi(b)\ Q^{\ell_\nu}(\gamma) \\ &= \chi(\gamma)\ Q^{\ell_\nu}(\gamma) \\ &= (\chi Q^{\ell_\nu})\ (\gamma) \end{aligned}$$

so that $Q^{\ell_{\chi\nu}} = \chi \cdot Q^{\ell_\nu}$. Since χ is arbitrary, the map (2.5.4) is surjective. But this relation and (2.5.5) also prove the remainder of the theorem. QED

2.6. Let (X,ω) be given with $[\omega]$ integral and let $[(L,\alpha)] = \ell \in \mathcal{L}_c(\omega)$ be fixed. We wish to determine the "Lie algebra" of $E(L,\alpha)$. Note of course that any element $\tau \in E(L,\alpha)$ leaves L^* invariant, and that conversely and diffeomorphism of L^* which (1) commutes with the action of $\underline{\underline{C}}^*$, (2) leaves α invariant and (3) leaves $|H|^2$ invariant where H is an α-invariant Hermitian structure in L , extends to a unique element of $E(L,\alpha)$. We may therefore regard $E(L,\alpha)$ as a group of diffeomorphisms of L^* .

Remark 2.6.1. The above is a special case of the fact that any diffeomorphism of a principal bundle which commutes with the action of the group of the bundle defines in a natural way a diffeomorphism of any associated bundle.

For any $x \in L^*$, let $\mathrm{Ver}_x(L)$ be the (2-dimensional) tangent space to $L^*_{\pi(x)}$ (or $L_{\pi(x)}$) at x and let

$$\mathrm{Hor}_x(L) = \mathrm{Ker}\,\alpha_x$$

where α_x is the value of α at x .

If τ is a diffeomorphism of a manifold then τ_* will denote its differential (a diffeomorphism of its tangent bundle).

Proposition 2.6.1. <u>For any</u> $x \in L^*$ <u>one has the direct sum</u>

$$T_x(L) = \mathrm{Ver}_x(L) \oplus \mathrm{Hor}_x(L) . \qquad (2.6.1)$$

<u>Moreover, if</u> τ <u>is a diffeomorphism of</u> L^* <u>which</u> (1) <u>commutes with the action of</u> $\underline{\underline{C}}^*$ <u>and satisfies</u> (2) $\tau^*\alpha = \alpha$ <u>then</u>

$$\tau_*(\mathrm{Ver}_x L) = \mathrm{Ver}_{\tau x} L \tag{2.6.2}$$

$$\tau_*(\mathrm{Hor}_x L) = \mathrm{Hor}_{\tau x} L \,. \tag{2.6.3}$$

Proof. Clearly $\mathrm{Ker}\ \alpha_x = \mathrm{Ker\ Re}\ \alpha_x \cap \mathrm{Ker\ Im}\ \alpha_x$. Thus $\mathrm{Ker}\ \alpha_x$ has codimension of at most 2 in $T_x(L)$. On the other hand since $\frac{1}{2\pi i}\frac{dz}{z}$ vanishes at no real tangent vector to $\underline{\underline{C}}^*$ it follows that $\mathrm{Ker}\ \alpha_x \cap \mathrm{Ver}_x(L) = 0$. This proves (2.6.1) . The relations (2.6.2) and (2.6.3) are obvious since $\tau(L^*_{\pi(x)}) = L^*_{\pi(\tau x)}$ and $\tau^*\alpha = \alpha$. QED

Now let $\underline{e} = \underline{e}(L)$ be the Lie algebra of all real vector fields η on L^* which "commute" with the action of $\underline{\underline{C}}^*$ on L^*, that is, which satisfy

$$c_*\eta_x = \eta_{cx} \tag{2.6.4}$$

for all $c \in \underline{\underline{C}}^*$, $x \in L^*$.

Also let

$$C(X) \longrightarrow C(L^*)$$

$\phi \longmapsto \tilde{\phi}$ be the injection given by $\tilde{\phi}(x) = \phi(\pi x)$. The image is denoted by $\tilde{C}$.

Remark 2.6.2. Note that from the local product structure on L^*, $\tilde{C}$ is the set of all $\phi \in C(L^*)$ which are invariant under the action of $\underline{\underline{C}}^*$ and hence by (2.6.5), C is stable under the action of $\underline{e}(L)$.

An element $\eta \in \underline{e}(L)$ is called vertical if $\eta_x \in Ver_x(L)$ and horizontal if $\eta_x \in Hor_x L$ for any $x \in L^*$.

Let ver L and hor L, respectively, be the spaces of all vertical and horizontal vector fields in $\underline{e}(L)$. It follows from Proposition 2.6.1 (with τ arising from an element in $\underline{\underline{C}}^*$) that as linear spaces

$$\underline{e}(L) = \text{ver }(L) \oplus \text{hor }(L) . \tag{2.6.5}$$

If M and N are manifolds and $\sigma: M \longrightarrow N$ is a smooth map then a vector field ξ on M and a vector field η on N are called σ-related in case $\sigma_* \xi_p = \eta_{\sigma p}$ for all $p \in M$. We can then write $\sigma_* \xi = \eta$. It is a simple fact that if ξ_i is σ-related to η_i, $i = 1,2$ then $[\xi_1, \xi_2]$ is σ-related to $[\eta_1, \eta_2]$ so that

$$\sigma_*[\xi_1, \xi_2] = [\sigma_* \xi_1, \sigma_* \xi_2] . \tag{2.6.6}$$

Now since $\tilde{\pi} \circ c = \tilde{\pi}$ where $\tilde{\pi}: L^* \longrightarrow X$ is the restriction of π and c is identified with the diffeomorphism it induces on L^*, one has by (2.6.4) the relation $(\tilde{\pi})_* \eta_x = (\tilde{\pi})_* \eta_y$ for any $\eta \in \underline{e}(L)$ where $\tilde{\pi}(x) = \tilde{\pi}(y)$. Thus given $\eta \in \underline{e}(L)$ one defines a vector field $\check{\eta}$ on X by the relation

$$\check{\eta}_p = \tilde{\pi}_*(\eta_x) \tag{2.6.7}$$

for any $x \in L_p^*$. To see that $\check{\eta}$ is smooth note that for any $\phi \in C(X)$ one has

$$\eta\tilde{\phi} = (\check{\eta}\phi)^{\sim}$$

is smooth and hence $\check{\eta}\phi$ is smooth (since local sections exist). Obviously η and $\check{\eta}$ are $\tilde{\pi}$-related so that by (2.6.6) one has

$$[\eta_1, \eta_2]^{\vee} = [\check{\eta}_1, \check{\eta}_2] \quad . \tag{2.6.8}$$

Remark 2.6.3. Note that ver(L) is the kernel of the Lie algebra homomorphism $\underline{e}(L,\alpha) \longrightarrow \underline{\mathfrak{v}}(X)$ given by $\eta \longrightarrow \check{\eta}$ and hence is an ideal in $\underline{e}(L)$.

2.7. A Lie group G will be said to operate smoothly on a manifold M if one has a homomorphism $\sigma: G \longrightarrow D(M)$ such that the map

$$G \times M \longrightarrow M \ , \qquad (g, p) \longrightarrow g \cdot p \ = \ \sigma(g)p \tag{2.7.1}$$

is smooth. In such a case $C(M)$ becomes a G-module where if $\phi \in C(M)$, $g \in G$, $p \in M$ one has

$$(g \cdot \phi)\ (p) \ = \ \phi(g^{-1} \cdot p) \quad . \tag{2.7.2}$$

If $G = \underline{\underline{R}}$ then σ is called a one-parameter group of diffeomorphisms.

A real vector field η on M is called globally integrable if there exists a one-parameter group of diffeomorphisms $\sigma(t)$, $t \in \underline{\underline{R}}$, such that for all $\phi \in C(M)$ one has

$$\begin{aligned} \eta(\phi)\ (p) \ &= \ \frac{d}{dt}\ \phi\big(\sigma(-t)\ p\big)\Big|_{t=0} \\ &= \ \frac{d}{dt}\ \big(\sigma(t) \cdot \phi\big)\ (p)\Big|_{t=0} \quad . \end{aligned} \tag{2.7.3}$$

In such a case one knows σ (called the generated one-parameter group) is unique and that conversely any σ defines η (called the infinitesmal generator of σ) by (2.7.3).

Now we recall (see 1.10.2) that any $\psi \in C^*$ defines a diffeomorphism τ_ψ of L^* where

$$\tau_\psi x = \tilde{\psi}(x) x \quad . \tag{2.7.4}$$

The elements of ver L are easy to describe.

Proposition 2.7.1. <u>For any</u> $\phi \in C(X)$ <u>there exists a unique real vector field</u> $\eta(\phi)$ <u>on</u> L^* <u>such that</u>

$$\big(\eta(\phi)\psi\big)(x) = \frac{d}{dt}\,\psi\left(e^{-2\pi i\, t\tilde{\phi}(x)}\psi\right)\Big|_{t=0} \tag{2.7.5}$$

<u>for any</u> $\psi \in C(L^*)$, $x \in L^*$. <u>Moreover</u> $\eta(\phi) \in$ <u>ver</u>(L) <u>and the map</u>

$$C(X) \longrightarrow \text{ver } L \ , \tag{2.7.6}$$

<u>given by</u> $\phi \longrightarrow \eta(\phi)$, <u>is a linear isomorphism</u>. <u>Also</u> $\eta(\phi)$ <u>is globally integrable for any</u> $\phi \in C$ <u>where the corresponding</u> 1-<u>parameter group is given by</u>

$$t \longrightarrow \tau_{\exp 2\pi i\, t\phi} \ .$$

<u>Proof</u>. It is obvious from (2.7.4) that $t \longrightarrow \tau_{\exp 2\pi i\, t\phi}$ defines a 1-parameter group of diffeomorphisms of L^* which commutes with the action of $\underline{\underline{C}}^*$ and hence defines an element $\eta(\phi) \in \underline{e}(L)$. Since the orbits of the 1-parameter group lie in the fibers L^*_p , $p \in X$ it follows that $\eta(\phi) \in \text{ver}(L)$.

Now let $\eta \in \text{ver}(L)$. For any $x \in L^*$ let $\psi_x \in C(L^*_{\pi(x)})$ be defined by $\psi_x(y) = \frac{y}{x}$ and put $\rho(x) = \frac{-1}{2\pi i} \eta_x \psi_x \in \underline{\underline{C}}^*$ defining a function ρ on L^* . But $c^*(\psi_{cx}) = \psi_x$ for any $c \in \underline{\underline{C}}^*$ (regarded as a diffeomorphism) so that (2.6.4) implies that ρ is $\underline{\underline{C}}^*$-invariant and hence there exists a unique function ϕ on X such that $\rho(x) = \phi(\pi x)$ for any $x \in L^*$. To see that ϕ is smooth note that if $s \in S^*(U)$ for some open set $U \subseteq X$ and $\psi_s \in C(\tilde{\pi}^{-1}U)$ is given by $\psi_s(y) = \frac{y}{s(\pi(y))}$ then

$$\phi \mid U = \eta\psi_s \circ s . \tag{2.7.7}$$

But then $\eta = \eta(\phi)$ since $\rho(x) = \frac{-1}{2\pi i} \eta(\phi)_x \psi_x$ by (2.7.5). This implies $(\eta(\phi))_x = \eta_x$ since no non-zero element of $\text{Ver}_x(L)$ can vanish on ψ_x . This proves that (2.7.6) is surjective. It is injective by (2.7.7) where $\eta = \eta(\phi)$. The last statement of the Proposition is obvious from the definition of $\eta(\phi)$. QED

Remark 2.7.2. Note that by Proposition 2.7.1 ver L is a commutative Lie algebra under Poisson bracket (since the group of all $\tau_\phi, \phi \in C^*(X)$, is clearly a commutative group).

The following corollary of the proof recovers ϕ from $\eta(\phi)$.

Proposition 2.7.2. <u>For any</u> $\phi \in C$ <u>one has</u>

$$\langle \alpha, \eta(\phi) \rangle = -\tilde{\phi} .$$

<u>Proof</u>. From the proof of Proposition 2.7.1 one has for any $x \in L^*$ the relation $\rho(x) = \frac{-1}{2\pi i} \eta_x \psi_x$. But where $\eta = \eta(\phi)$ this implies

$$\tilde{\phi}(x) = \frac{-1}{2\pi i} \eta(\phi)_x \psi_x . \tag{2.7.8}$$

But clearly $\alpha | L^*_x = \frac{1}{2\pi i} \frac{d\psi_x}{\psi_x}$. However $\psi_x(x) = 1$ so that $\langle \alpha, \eta(\phi) \rangle (x) = \frac{1}{2\pi i} \eta(\phi)_x \psi_x$. Applying (2.7.8) proves the Proposition.

2.8. Let $\underline{d}(X)$ denote the Lie algebra of all real vector fields on X $(\mathfrak{v} = \underline{d} + i\underline{d})$.

Proposition 2.8.1. <u>For any</u> $\xi \in \underline{d}(X)$ <u>there exists a unique</u> $\tilde{\xi} \in \text{hor}(L)$ <u>such that</u> ξ <u>and</u> $\tilde{\xi}$ <u>are</u> $\tilde{\pi}$<u>-related. Moreover the map</u>

$$\underline{d}(X) \longrightarrow \text{hor}(L) \tag{2.8.1}$$

<u>is a linear isomorphism.</u>

<u>Proof</u>. From the local product structure in L^* one has, for any $x \in L^*$, an exact sequence

$$0 \longrightarrow \text{Ver}_x(L) \longrightarrow T_x(L) \xrightarrow{\pi_*} T_{\pi(x)}(X) \longrightarrow 0 .$$

But then $\pi_* : \text{Hor}_x(L) \longrightarrow T_{\pi(x)}(X)$ is an isomorphism by (2.6.1) so that if $\tilde{\xi} \in \text{hor}(L)$ exists such that ξ and $\tilde{\xi}$ are $\tilde{\pi}$-related, $\tilde{\xi}$ is unique. In fact to show the map (2.8.1) exists we have only to show that if $\tilde{\xi}$ is the vector field on L^* given by putting $\tilde{\xi}_x$ equal to the unique element of $\text{Hor}_x(L)$ such that $\pi_*(\tilde{\xi}_x) = \xi_x$ then $\tilde{\xi}$ is smooth. (It is clearly $\underline{\underline{C}}^*$-invariant.)

Let $U \subseteq X$ be open such that $S^*(U)$ is not empty. If $s \in S^*(U)$ let ξ_s be the real vector field on $\underline{\underline{C}}^* \times U$ given by

$$(2.8.2) \qquad (\xi_s)_{(c,p)} = \left[2\pi i\left(\langle \overline{\alpha(s)}_p, \xi_p\rangle\, \bar{c}\left(\frac{\partial}{\partial \bar{z}}\right)_c - \langle \alpha(s)_p, \xi_p\rangle c\left(\frac{\partial}{\partial z}\right)_c\right), \xi_p\right].$$

Clearly ξ_s is real and smooth and one has $\langle \sigma_s^* \alpha, \xi_s\rangle = 0$ by (1.5.2) since $\frac{\partial}{\partial \bar{z}}$ is orthogonal to dz. But since $z\frac{\partial}{\partial z}$ and $\bar{z}\frac{\partial}{\partial \bar{z}}$ are $\underline{\underline{C}}^*$ invariant, it follows that $(\sigma_s)_* \xi_s \in$ hor $\tilde{\pi}^{-1}(U)$. However if $x = \sigma_s(c,p)$ then $\pi_*\left((\sigma_s)_* \xi_s\right)_x = \xi_p$ so that we must have

$$(\sigma_s)_* \xi_s = \tilde{\xi}|\tilde{\pi}^{-1}(U) . \tag{2.8.3}$$

This proves that $\tilde{\xi}$ is smooth.

The map (2.8.1) is obviously injective but it is also surjective since if $\eta \in$ hor L then $\pi_* \eta_x = \pi_* \eta_{cx}$ for any $c \in \underline{\underline{C}}^*$, $x \in L^*$ by (2.6.4) and hence there exists a unique vector field ξ on X such that for any $x \in L^*$ one has $\pi_* \eta_x = \xi_{\pi x}$. But ξ is smooth since, for any $\phi \in C$, $\widetilde{\xi\phi} = \eta\tilde{\phi}$ which implies that $\widetilde{\xi\phi}$ and hence $\xi\phi$ is smooth. Obviously $\eta = \tilde{\xi}$. QED

2.9. One now has that $C(X) \times \underline{d}(X)$ parameterizes $\underline{e}(L)$. That is for any $\phi \in C$, $\xi \in \underline{d}$ let $\eta(\phi,\xi) = \eta(\phi) + \tilde{\xi}$ so that $\eta(\phi,\xi) \in \underline{e}(L)$.

Proposition 2.9.1. _The map_

$$C \times \underline{d} \longrightarrow \underline{e}(L)$$

given by $(\phi,\xi) \longrightarrow \eta(\phi,\xi)$ _is a linear isomorphism. Moreover if_ H _is an_ α-_invariant Hermitian structure in_ L, _then_

$$\eta(\phi,\xi)\ |H|^2 = 0$$

if and only if ϕ *is real.*

Proof. The first statement follows from (2.6.5) and Propositions 2.7.1 and 2.8.1. But now by (1.9.2) $\frac{d|H|^2}{|H|^2} = 2\pi i\,(\alpha - \bar{\alpha})$.
However if $\xi \in \underline{d}$ then $\langle \alpha, \tilde{\xi} \rangle = 0$ and hence $\langle \bar{\alpha}, \tilde{\xi} \rangle = 0$ since ξ is real. Thus $\langle d\,|H|^2, \tilde{\xi} \rangle = \tilde{\xi}\,|H|^2 = 0$. But since $\eta(\phi)$ is real $\langle \alpha - \bar{\alpha}, \eta(\phi) \rangle = (\bar{\phi} - \phi)$ by Proposition 2.7.2. Thus $\langle d\,|H|^2, \eta(\phi) \rangle = \eta(\phi)\,|H|^2 = 0$ if and only if ϕ is real. QED

2.10. Now if ξ is a real vector field on a manifold M then one knows that ξ is globally integrable if for any $p \in M$ there exists a global trajectory γ for ξ through p. That is a smooth curve $\gamma : \underline{\underline{R}} \longrightarrow M$ such that $\gamma(0) = p$ and such that $\gamma'(t) = \xi_{\gamma(t)}$ for all t.

Note of course that a curve $\tilde{\gamma}(t)$ in L^* is a nowhere vanishing section along the curve $\gamma(t) = \pi(\tilde{\gamma}(t))$ in X. The following is a generalization of Proposition 2.7.1.

Theorem 2.10.1. *Let* $\eta \in \underline{e}(L)$ *so that* $\eta = \eta(\phi, \xi)$ *for a unique* $\phi \in C(X)$, $\xi \in \underline{d}(X)$. *Then* η *is globally integrable on* L^* *if and only if* ξ *is globally integrable in* X. *Moreover in such a case if* $x \in L^*$ *and* $p = \pi(x) \in X$ *then the global trajectory of* η *through* x *is the curve* $\tilde{\gamma}$ *given by*

$$\tilde{\gamma}(t) = g(t)\, s(t) \tag{2.10.1}$$

where γ *is the global trajectory of* ξ *through* p, s *is the*

<u>unique auto-parallel section along</u> γ <u>such that</u> $s(0) = x$ <u>and</u> g <u>is the function on</u> $\underline{\underline{R}}$ <u>given by</u>

$$g(t) = e^{-2\pi i \int_0^t \phi(\gamma(r))\, d r} \quad . \tag{2.10.2}$$

<u>Proof</u>. Since $\pi_*(\eta(\phi,\xi)_y) = \xi_{\pi y}$, it follows that if $\tilde{\gamma}$ is a global trajectory of $\eta(\phi,\xi)$ through x then $\gamma = \pi \circ \tilde{\gamma}$ is a global trajectory for ξ through $p = \pi x$. Hence ξ is globally integrable in case η is globally integrable.

Now assume ξ is globally integrable and let γ be the global trajectory of ξ through $p = \pi x$. Let s be the auto-parallel section along γ such that $s(p) = x$. Then for any $t \in \underline{\underline{R}}$

$$s'(t) \in \mathrm{Hor}_{s(t)} L$$

since $\langle \alpha, s' \rangle = 0$ by (1.7.2). But since $\langle \alpha, s' \rangle = \langle s^*\alpha, \frac{d}{dt} \rangle$ this also implies that $s^*(\alpha) = 0$. Hence if g is any non-vanishing smooth function on $\underline{\underline{R}}$ and we put $r = g\, s$ then by (1.5.3) one has

$$r^*(\alpha) = \frac{1}{2\pi i} \frac{dg}{g} \quad .$$

Thus $\langle \alpha, r' \rangle = \langle r^*\alpha, \frac{d}{dt} \rangle = \frac{1}{2\pi i} \frac{1}{g} \frac{dg}{dt}$. But then if we let g be given by (2.10.2) we have

$$\langle \alpha, r' \rangle = -\phi \circ \gamma \; . \tag{2.10.3}$$

On the other hand if we let $\eta(\phi) \circ r$ be a vertical vector field along γ given by $(\eta(\phi) \circ r)\,(t) = \eta(\phi)_{r(t)}$ one has

$$\langle \alpha, \eta(\phi) \circ r \rangle = -\phi \circ \gamma$$

by Proposition 2.7.2 since $\tilde{\phi}(r(t)) = \phi\Big[\pi(r(t))\Big] = \phi(\gamma(t))$.
But then by (2.10.3) if $h = r' - \eta(\phi) \circ r$, h is a horizontal vector field along γ and hence $r' = (\eta(\phi) \circ r) + h$ is the unique decomposition of r' into its vertical and horizontal components. On the other hand $\pi_* r'(t) = \gamma'(t) = \xi_{\gamma(t)}$. Thus the horizontal component of $r'(t)$ is $\tilde{\xi}_{r(t)}$. That is $h(t) = \tilde{\xi}_{r(t)}$. Thus for all $t \in \underline{\underline{R}}$ one has

$$r'(t) = \eta(\phi)_{r(t)} + \tilde{\xi}_{r(t)} = \big(\eta(\phi,\xi)\big)_{r(t)} \ . \tag{2.10.4}$$

Since $r(0) = x$ this proves that $\tilde{\gamma} = r$ is the global trajectory of $\eta(\phi,\xi)$ through x . QED

3. The Lie algebra of connection preserving vector fields as a central extension of the Lie algebra of Hamiltonian vector fields.

1. Now for any vector field $\xi \in \underline{\mathfrak{n}} = \underline{d} + i\,\underline{d}$ on X let β_ξ be the 1-form on X given by

$$\beta_\xi = i\,(\xi)\omega$$

where $\omega = \text{curv}\,(L,\alpha)$. That is if $\eta \in \underline{\mathfrak{n}}$ one has

$$\langle \beta_\xi, \eta \rangle = \omega(\xi,\eta) \ . \tag{3.1.1}$$

Concerning the Lie derivative of ω by ξ one has

Proposition 3.1.1. For any $\xi \in \underline{\mathfrak{n}}$ one has

$$\theta(\xi)\omega = d\beta_\xi$$

Proposition 3.2.1. <u>If</u> $\xi,\eta \in \mathfrak{n}$ <u>where</u> $\xi \in \underline{\mathfrak{n}}(X,\omega)$ <u>then</u>

$$\theta(\xi)\beta_\eta = \beta_{[\xi,\eta]} \cdot \qquad (3.2.1)$$

<u>Also</u> $\mathfrak{a}$ <u>contains the cummutator of</u> $\underline{d}(X,\omega)$ <u>with itself so that in particular</u> $\mathfrak{a}$ <u>is an ideal in</u> $\underline{d}(X,\omega)$.

<u>Proof</u>. For any $\xi,\eta \in \mathfrak{n}$ one knows that (see (1.3.2)) $[\theta(\xi), i(\eta)] = i[(\xi,\eta)]$. Now if $\xi \in \underline{\mathfrak{n}}(X,\omega)$ then one obtains (3.2.1) by applying $\theta(\xi)$ to the relation $i(\eta)\omega = \beta_\eta$ since $\theta(\xi)\omega = 0$.

But now if both $\xi,\eta \in \underline{\mathfrak{n}}(X,\omega)$ then by the relation (1.3.1) one has $\beta_{[\xi,\eta]} = \theta(\xi)\beta_\eta = d\, i(\xi)\beta_\eta$ so that $\beta_{[\xi,\eta]}$ is exact. If ξ and η are real so is the function $i(\xi)\beta_\eta$. Hence $[\xi,\eta] \in \underline{\mathfrak{a}}(X,\omega)$. QED

Remark 3.2.2. Note that if $\xi,\eta \in \underline{\mathfrak{n}}(X,\omega)$ and $\rho \in \underline{\mathfrak{n}}$ is arbitrary one has

$$\omega([\xi,\eta], \rho) = \rho\omega(\eta,\xi) . \qquad (3.2.2)$$

Indeed the left side is $\langle \beta_{[\xi,\eta]}, \rho \rangle$ whereas the right side is $\langle d(\omega(\eta,\xi)), \rho \rangle$. But $\omega(\eta,\xi) = \langle \beta_\eta, \xi \rangle = i(\xi)\beta_\eta$. However $\beta_{[\xi,\eta]} = d\, i(\xi)\beta_\eta$ from the the proof above. This establishes (3.2.2) and also the relation

$$\beta_{[\xi,\eta]} = d(\omega(\eta,\xi)) . \qquad (3.2.3)$$

3.3. Now assume $\ell = [(L,\alpha)] \in \mathcal{L}_c(X,\omega)$. We consider the "Lie algebra" $\underline{e}(L,\alpha)$ of the group $E(L,\alpha)$ (see § 1.12). Let $\underline{e}(L,\alpha)$ be the set of all $\eta \in \underline{e}(L)$ such that $\theta(\eta)\alpha = 0$ and $\eta|H|^2 = 0$ where H is an α-invariant Hermitian structure in L.

Remark 3.3.1. It is clear that if $\eta \in \underline{e}(L)$ is globally integrable and σ is the corresponding one-parameter group, then $\eta \in \underline{e}(L,\alpha)$ if and only if $\sigma(t) \in E(L,\alpha)$ for all t .

For any $\eta \in \underline{e}(L)$ one determines $\theta(\eta)\alpha$ by

Lemma 3.3. Let $\phi \in C(X)$ and $\xi \in \underline{d}(X)$ then

$$\theta\big(\eta(\phi,\xi)\big)\alpha = \tilde{\pi}^*(\beta_\xi - d\phi) .$$

Proof. Let $\eta = \eta(\phi,\xi)$ so that $\eta = \eta(\phi) + \tilde{\xi}$. Now $\theta(\eta) = i(\eta)\, d + d\, i(\eta)$. But $i(\eta)\alpha = \langle\alpha,\eta\rangle = -\tilde{\phi}$ by Proposition 2.7.2. Hence

$$d\, i(\eta)\alpha = -d\tilde{\phi} = \tilde{\pi}^*(-d\phi) .$$

On the other hand $d\alpha = \tilde{\pi}^*\omega$ so that

$$\begin{aligned} i(\eta)\, d\alpha &= i(\eta)\, \tilde{\pi}^*\omega \\ &= \tilde{\pi}^*\, i(\xi)\omega \\ &= \tilde{\pi}^*\beta_\xi \end{aligned}$$

since $\check{\eta} = \xi$. QED

We can explicity give $\underline{e}(L,\alpha)$. The following also contains the infinitesmal version of Theorem 1.13.1. It is more constructive than Theorem 1.13.1 since $\underline{e}(L,\alpha)$ is explicitly given.

Theorem 3.3.1. *Let* $\eta \in \underline{e}(L)$ *so that* $\eta = \eta(\phi,\xi)$ *for* $\phi \in C(X)$ *and* $\xi \in \underline{d}(X)$. *Then* $\eta \in \underline{e}(L,\alpha)$ *if and only if* (1) ϕ *is real and* (2)

$$\beta_\xi = d\phi \quad .$$

Moreover one has an exact sequence of Lie algebras

$$0 \longrightarrow \underline{\underline{R}} \longrightarrow \underline{e}(L,\alpha) \longrightarrow \underline{\mathfrak{a}} \longrightarrow 0 \tag{3.3.1}$$

giving $\underline{e}(L,\alpha)$ *as a central extension of* $\underline{\mathfrak{a}}$ *by* $\underline{\underline{R}}$. *Here the map* $\underline{e}(L,\alpha) \longrightarrow \underline{\mathfrak{a}}$ *is defined by the correspondence* (*see* 2.6.7) $\eta \longrightarrow \check{\eta}$ *and* $\underline{\underline{R}} \longrightarrow \underline{e}(L,\alpha)$ *is defined by* $r \longrightarrow \eta(r)$ *where* r *is identified with the constant function of* X *with value* r .

Proof. By Proposition 2.9.1. one has $\eta(\phi,\xi)\,|H|^2 = 0$ if and only if ϕ is real. On the other hand since $\tilde{\pi}^*$ is injective one has $\theta(\eta)\alpha = 0$ if and only if $\beta_\xi = d\phi$ by Lemma 3.3.

But now if $\eta = \eta(\phi,\xi) \in \underline{e}(L,\alpha)$ then $\check{\eta} = \xi \in \underline{\mathfrak{a}}$ since $\beta_\xi = d\phi$ is exact. The map $\underline{e}(L,\alpha) \longrightarrow \underline{\mathfrak{a}}$ is clearly surjective since if $\xi \in \underline{\mathfrak{a}}$ there exists a real function ϕ such that $\beta_\xi = d\phi$. Hence $\check{\eta} = \xi$ where $\eta = \eta(\phi,\xi)$. But ϕ is unique up to a real constant proving the exactness of the sequence. The maps are Lie algebra homomorphisms by (2.6.8). Also the image $\underline{\underline{R}}$ is obviously central by definition of $\underline{e}(L)$. QED

Corollary to Theorem 3.3.1. If $\xi \in \mathfrak{a}$ is globally integrable and σ is the corresponding 1-parameter group of diffeomorphisms then for any $\ell \in \mathcal{L}_c(X,\omega)$ one has $\sigma(t) \in D_\ell(X)$ for all $t \in \underline{\underline{R}}$ (see § 1.13).

Proof. Let $[(L,\alpha)] = \ell \in \mathcal{L}_c(\omega)$. Now let $\phi \in C(X)$ be real such that $d\phi = \beta_\xi$ and let $\eta = \eta(\phi,\xi)$. Then $\check{\eta} = \xi$ and η is globally integrable by Theorem 2.10.1. If σ_η is the corresponding 1-parameter group then clearly $\check{\sigma}_\eta(t) = \sigma(t)$. But $\sigma_\eta(t) \in E(L,\alpha)$ since $\eta \in \underline{e}(L,\alpha)$ and hence $\sigma(t) \in D_\ell(\omega)$ by Theorem 1.13.1. QED

Remark 3.3.2. One has a natural map $H^1(X, \underline{\underline{R}}) \to \Pi^*$, $\mu \longrightarrow \chi_\mu$ where χ_μ is given by $\chi_\mu(\gamma) = e^{2\pi i \int_\gamma \beta}$ for any closed 1-form β such that $[\beta] = \mu$. We will not use the fact but it can be easily shown that the corollary above generalizes to the following: if the above ξ is just in $\underline{d}(X,\omega)$ instead of $\mathfrak{a}$ so that β_ξ is only closed and not necessarily exact, one has

$$\sigma(t) \cdot \ell = \chi_{t[\beta_\xi]}\, \ell$$

(see Theorem 2.5.1) where $[\beta_\xi] \in H^1(X, \underline{\underline{R}})$ is the de Rham class of β_ξ .

3.4. Now we recall that S_m is the space of all measurable sections of L . The space is a module for the group $E(L)$ where for any $\tau \in E(L)$, $s \in S_m$, $p \in X$ one puts

$$(\tau \cdot s)(p) = \tau\big(s(\tau^{-1} \cdot p)\big)$$

where for convenience we write $\tau \cdot p$ for $\check{\tau} \cdot p$.

This action may be reduced to an action on ordinary functions as follows: Let $\tilde{S}_m$ be the space of all measurable complex-valued functions u on L^* such that

$$c \cdot u = cu \tag{3.4.1}$$

for all $c \in \underline{\underline{C}}^*$. That is all u such that for any $x \in L^*$, $u(c^{-1} x) = c\, u(x)$. Since $E(L)$ commutes with the action of $\underline{\underline{C}}^*$ it is clear that $\tilde{S}_m$ is stable under the action of $E(L)$.

Proposition 3.4.1. <u>For any</u> $s \in S_m$ <u>let</u> $\tilde{s}$ <u>be the function on</u> L^* <u>defined by</u>

$$\tilde{s}(x) = \frac{s(\pi x)}{x} . \tag{3.4.2}$$

<u>Then</u> $\tilde{s} \in S_m$ <u>and the map</u>

$$S_m \longrightarrow \tilde{S}_m \tag{3.4.3}$$

<u>defined by</u> $s \longrightarrow \tilde{s}$ <u>is an</u> $E(L)$ <u>isomorphism</u>.

<u>Proof</u>. It is clear from (3.4.2) that $\tilde{s}(c^{-1} x) = c\, \tilde{s}(x)$. Also if $r \in S^*(U)$ for some open $U \subseteq X$ then

$$(\tilde{s} \circ r)\, r = s$$

and hence (see (1.5.1))

$$(\tilde{s} \circ \sigma_r)\,(c,p) = c^{-1} \frac{s}{r}\,(p) \tag{3.4.4}$$

which proves that $\tilde{s}$ is measurable so that $\tilde{s} \in S_m$. Also obviously $\tilde{s} = 0$ implies $s = 0$. On the other hand $u \in \tilde{S}_m$ defines $s \in S_m$ such that $\tilde{s} = u$ where, on U, $(u \circ r)\, r = s$. Since $(u \circ gr)\, gr = (u \circ r)\, r$ for $g \in C^*(U)$ by (3.4.1) this is independent of the choice of r proving that the map is an isomorphism.

Now if $\tau \in E(L)$, $s \in S_m$, $x \in L^*$ then $(\tau \cdot \tilde{s})(x) = \tilde{s}(\tau^{-1} x) = \frac{s(\tau^{-1} p)}{\tau^{-1} x}$ where $p = \pi x$. But since τ is linear on $L_{\tau^{-1}p}$ one has $\frac{s(\tau^{-1}p)}{\tau^{-1}x} = \frac{\tau\left(s(\tau^{-1} p)\right)}{x} = \frac{(\tau \cdot s)(p)}{x} = \widetilde{\tau \cdot s}(x)$.

Hence $\tau \cdot \tilde{s} = \widetilde{\tau \cdot s}$. QED

Let $\tilde{S}$ be the image of S under the map (3.4.3). Then by (3.4.4) clearly $\tilde{S}$ is the set of all smooth functions in $\tilde{S}_m$. It is also clearly stable under $E(L)$ and from the relations

$$\begin{aligned} c \cdot (\eta\tilde{s}) &= (c_* \eta)\; c \cdot \tilde{s} \\ &= \eta\; c\tilde{s} \\ &= c\; \eta\tilde{s} \end{aligned}$$

where $c \in \underline{\underline{C}}^*$ note that $\tilde{S}$ is stable under $\underline{e}(L)$. The corresponding action on S is given by

Proposition 3.4.2. <u>For any real</u> $\phi \in C$, $\xi \in \underline{d}$ <u>and</u> $s \in S$ <u>one has</u> $\eta(\phi,\xi)\tilde{s} = \tilde{t}$ <u>where</u> $t \in S$ <u>is given by</u>

$$t = (\nabla_\xi + 2\pi i\, \phi)\, s$$

(ϕ <u>being regarded here as the multiplication operator by the function</u> ϕ).

Proof. By definition

$$\left(\eta(\phi)\tilde{s}\right)(x) = \frac{d}{dt}\,\tilde{s}(e^{-2\pi i\, t\tilde{\phi}(x)}x)\Big|_{t=0} .$$

But $\tilde{s}(e^{-2\pi i\, t\tilde{\phi}(x)}x) = e^{2\pi i\, t\tilde{\phi}(x)}\tilde{s}(x)$. Hence

$$\eta(\phi)\tilde{s} = 2\pi i\ \tilde{\phi}\tilde{s}$$

$$= 2\pi i\ \widetilde{\phi s}$$

since clearly

$$\widetilde{\phi s} = \tilde{\phi}\tilde{s} . \tag{3.4.5}$$

We have only to prove that $\tilde{\xi}\tilde{s} = \widetilde{\nabla_\xi s}$. Since this is a local relation it is enough to prove this for any element $s \in S^*(U)$ and any open $U \subseteq X$. Using (1.4.6), (3.4.5) and the relation $\tilde{\xi}\tilde{\psi} = \widetilde{\xi\psi}$, for $\psi \in C$, one then proves it for a general local section. But now by (1.5.1) $s \circ \sigma_s$ as a function on $\underline{\underline{C}}^* \times U$ is given by $(\tilde{s} \circ \sigma_s)(c,p) = c^{-1}$. On the other hand if ξ_s is the vector field on $\underline{\underline{C}}^* \times U$ given by (2.8.3) then

$$\left(\xi_s(\tilde{s} \circ \sigma_s)\right)(c,p) = 2\pi i \left\langle \alpha(s)_p, \xi_p \right\rangle c^{-1}$$

$$= 2\pi i \left\langle \alpha(s)_p, \xi_p \right\rangle \left((s \circ \sigma_s)\ (c,p)\right) .$$

But $(\sigma_s)_* \xi_s = \tilde{\xi}$ by (2.8.3). Hence in $\tilde{\pi}^{-1}(U)$ one has

$$\tilde{\xi}\tilde{s} = 2\pi i \left\langle \alpha(\tilde{s}),\xi \right\rangle \tilde{s} = 2\pi i \widetilde{\left(\left\langle \alpha(s),\xi \right\rangle s\right)}$$

$$= (\widetilde{\nabla_\xi s}) .$$

QED

The bracket relations in $\underline{e}(L)$ are given in

Proposition (3.4.3). *Let* $\phi_i \in C(X)$ *and* $\xi_i \in \underline{d}(X)$ *for* $i = 1,2$. *Then*

$$[\eta(\phi_1,\xi_1), \eta(\phi_2,\xi_2)] = \eta(\xi_1\phi_2 - \xi_2\phi_1 + \omega(\xi_1,\xi_2), [\xi_1,\xi_2]) .$$

Proof. Let $\eta_i = \eta(\phi_i,\xi_i)$ and let $[\eta_1,\eta_2] = \eta(\phi,\xi) = \eta$. Clearly $\xi = [\xi_1,\xi_2]$ by (2.6.8) since $\check{\eta}_i = \xi_i$ and $\check{\eta} = \xi$. On the other hand

$$\langle\alpha, [\eta_1,\eta_2]\rangle = -\tilde{\phi} . \tag{3.4.6}$$

by Proposition 2.7.2.

But now by $\tilde{\pi}^*\omega = d\alpha$ and hence $d\alpha(\eta_1,\eta_2) = \omega(\tilde{\xi}_1,\tilde{\xi}_2)$. Thus by (1.3.6) one has

$$-\langle\alpha, [\eta_1,\eta_2]\rangle = \omega(\tilde{\xi}_1,\tilde{\xi}_2) - \eta_1\langle\alpha,\eta_2\rangle + \eta_2\langle\alpha,\eta_1\rangle .$$

But $\tilde{\phi}_2 = -\langle\alpha,\eta_2\rangle$ and $\tilde{\phi}_1 = -\langle\alpha,\eta_1\rangle$ and hence by (3.4.6)

$$\tilde{\phi} = \omega(\widetilde{\xi_1,\xi_2}) + \xi_1\tilde{\phi}_2 - \xi_2\tilde{\phi}_1 .$$

This proves the proposition. QED

With regard to the action on S the significant part of Proposition 3.4.3 yields

Proposition 3.4.4. *Let* $\xi_i \in \underline{d}$, $i = 1,2$ *and let* $s \in S$ *then*

$$\left(\left[\nabla_{\xi_1}, \nabla_{\xi_2}\right] - \nabla_{[\xi_1,\xi_2]}\right) s = 2\pi i\, \omega(\xi_1,\xi_2)\, s .$$

Proof. We choose $\phi_1 = \phi_2 = 0$ in Proposition (3.4.3) so that Proposition 3.4.3 asserts that

$$[\tilde{\xi}_1,\tilde{\xi}_2] = \eta\big(\omega(\xi_1,\xi_2)\ ,\ [\xi_1,\xi_2]\big)$$

$$= [\xi_1,\xi_2]^{\sim} + \eta\big(\omega(\xi_1,\xi_2)\big)$$

or

$$[\tilde{\xi}_1,\tilde{\xi}_2] - [\xi_1,\xi_2]^{\sim} = \eta\big(\omega(\xi_1,\xi_2)\big)\ . \tag{3.4.7}$$

Now apply (3.4.7) to $\tilde{s}$ and the result follows from Proposition 3.4.2.

QED

The bracket expression in Proposition 3.4.3 above simplifies considerably if $\eta_1 \in \underline{e}(L,\alpha)$. Indeed if $\beta_\xi = d\phi$ for $\phi \in C$ and $\xi \in \underline{d}$ one has for any $\eta \in \underline{n}$ that

$$\eta\phi = \omega(\xi,\eta) \tag{3.4.8}$$

since $\eta\phi = \langle d\phi,\eta\rangle = \langle i(\xi)\omega,\eta\rangle = \omega(\xi,\eta)$.

Proposition 3.4.5. With the notation of Proposition 3.4.3 if $\eta(\phi_1,\xi_1) \in \underline{e}(L,\alpha)$ then

$$[\eta(\phi_1,\xi_1),\ \eta(\phi_2,\xi_2)] = \eta(\xi_1\phi_2,\ [\xi_1,\xi_2])\ . \tag{3.4.9}$$

Proof. By (3.4.8) and Theorem 3.3.1 one has $\xi_2\phi_1 = \omega(\xi_1,\xi_2)$ yielding the simplication above.

QED

Remark 3.4.1. Note that if also $\eta(\phi_2,\xi_2) \in \underline{e}(L,\alpha)$ then the right side of (3.4.9) is an $\underline{e}(L,\alpha)$ so that $d\xi_1\phi_2 = \beta_{[\xi_1,\xi_2]}$.

But this is clear from (3.2.3) since $\xi_1\phi_2 = \omega(\xi_2\xi_1)$ by (3.4.8).

Remark 3.4.2. Hitherto we have been assuming that $[\omega]$ is integral. The central extension $\underline{e}(L,\alpha)$ of $\underline{\mathfrak{a}}$ that an element $\ell \in \mathcal{L}_c(X,\omega)$ provides can however be defined without assuming that $[\omega]$ is integral (so that $\mathcal{L}_c(X,\omega)$ may be empty). Indeed by Proposition 3.4.5 we need only substitute for $\underline{e}(L,\alpha)$ the set $\underline{\mathfrak{a}}^e$ of all pairs (ϕ,ξ) where ϕ is real, and $\xi \in \underline{d}$ is such that $d\phi = \beta_\xi = i(\xi)\omega$. One then defines

$$[(\phi_1,\xi_1), (\phi_2,\xi_2)] = (\xi_1\phi_2, [\xi_1,\xi_2]) .$$

This is a Lie algebra. The map $\underline{\mathfrak{a}}^e \longrightarrow \underline{\mathfrak{a}}$ given by $(\phi,\xi) \longrightarrow \xi$ defines $\underline{\mathfrak{a}}^e$ as a central extension of $\underline{\mathfrak{a}}$ by $\underline{R}$. We now observe that in the symplectic case this extension occurs in a far more natural way.

4. The symplectic case and prequantization.

The pair (X,ω) define the structure of a symplectic manifold X in case the alternating 2-form ω_p on $T_p(X)$ is non-singular for all $p \in X$. In this case we say simply that (X,ω) is symplectic.

Assume (X,ω) is symplectic but that $[\omega]$ is not necessarily integral. The correspondence $\xi \dashrightarrow \beta_\xi$ now defines an isomorphism

$$\underline{\mathfrak{U}}(X) \longrightarrow \Omega^1(X) \tag{4.1.1}$$

where $\underline{d}(X)$ corresponds to the set of all real 1-forms, $\underline{\mathfrak{U}}(X,\omega)$ (respectively $\underline{d}(X,\omega)$) corresponds to the set of all (respectively, real) closed 1-forms and $\underline{\mathfrak{a}}$ corresponds to the set of all real exact 1-forms. To every function $\phi \in C$ we now define an element ξ_ϕ in the complexification $\underline{\mathfrak{a}}_{\underline{\underline{C}}}$ of $\underline{\mathfrak{a}}$ defined by putting ξ_ϕ equal to the unique vector field (since (4.1.1) is an isomorphism) on X such that

$$i(\xi_\phi)\omega = \beta_{\xi_\phi} = d\phi \ . \tag{4.1.2}$$

The relation (3.4.8) now takes the following form: For any $\psi \in C$ and $\eta \in \underline{\mathfrak{v}}$ one has

$$\eta\psi = \omega(\xi_\psi, \eta) \ . \tag{4.1.3}$$

It is clear that if $R = R(X)$ denotes the space of all real-valued smooth functions on X then one has an exact sequence

$$0 \longrightarrow \underline{\underline{R}} \longrightarrow R \longrightarrow \underline{\mathfrak{a}} \longrightarrow 0 \tag{4.1.4}$$

where $\underline{\underline{R}} \longrightarrow R$ is injection and $R \longrightarrow \underline{\mathfrak{a}}$ is the map given by $\phi \longrightarrow \xi_\phi$.

One now defines a bracket operation (Poisson bracket of functions) in C by putting

$$[\phi, \psi] = \xi_\phi \psi \ . \tag{4.1.5}$$

Proposition 4.1.1. *The space* $C = C(X)$ *is a Lie algebra under the bracket operation* (4.1.5). *Moreover the map* $C \longrightarrow \underline{\mathfrak{a}}_{\underline{\underline{C}}}$, $\phi \longrightarrow \xi_\phi$, *is a homomorphism. Also* R *is a real form of* C *and the sequence* (4.1.4) *is an exact sequence of Lie algebras defining* R *as a central extension of* $\underline{\mathfrak{a}}$ *by* $\underline{\underline{R}}$.

<u>Proof</u>. By (4.1.3) one has $\xi_\phi \psi = \omega(\xi_\psi, \xi_\phi)$ so that

$$[\phi,\psi] = \omega(\xi_\psi, \xi_\phi) \tag{4.1.6}$$

which implies that $[\phi,\psi] = -[\psi,\phi]$. But now since $\xi_\psi, \xi_\phi \in \underline{\mathfrak{v}}(X,\omega)$ then $\beta_{[\xi_\phi,\xi_\psi]} = d(\omega(\xi_\psi,\xi_\phi))$ by (3.2.3). Hence $\beta_{[\xi_\phi,\xi_\psi]} = d[\phi,\psi]$. But $d[\phi,\psi] = \beta_{\xi_{[\phi,\psi]}}$ by definition of $\xi_{[\phi,\psi]}$. Thus

$$[\xi_\phi,\xi_\psi] = \xi_{[\phi,\psi]} \tag{4.1.7}$$

since (4.1.1) is an isomorphism.

Now let $\zeta \in C$ then since ω is closed, one has by (1.3.8)

$$\sum_{\substack{\text{cyclic}\\ \zeta,\phi,\psi}} \left(\xi_\zeta\, \omega(\xi_\psi,\xi_\phi) - \omega([\xi_\psi,\xi_\phi],\xi_\zeta)\right) = 0 .$$

On the other hand by (3.2.3) one has $\xi_\zeta\, \omega(\xi_\psi,\xi_\phi) = -\omega([\xi_\psi,\xi_\phi],\xi_\zeta)$. Thus one has

$$\sum_{\substack{\text{cyclic}\\ \zeta,\phi,\psi}} \xi_\zeta\, \omega(\xi_\psi,\xi_\phi) = 0 . \tag{4.1.8}$$

But $\xi_\zeta(\omega(\xi_\psi,\xi_\phi)) = \xi_\zeta[\phi,\psi] = [\zeta,[\phi,\psi]]$. Thus the Jacobi identity is satisfied. Hence C is a Lie algebra and by (4.1.7) the map $\phi \longrightarrow \xi_\phi$ is a Lie algebra homomorphism. Clearly $[\phi,\psi] \in R$ if $\phi,\psi \in R$. Also the image of $\underline{\underline{R}}$ of R is obviously central in R.

QED

4.2. Now assume $[\omega]$ is integral and let $\ell = [(L,\alpha)] \in \mathcal{L}_c(\omega)$. We now observe that R is equivalent to $\underline{e}(L,\alpha)$ as a central extension of $\underline{\mathfrak{a}}$. For any $\phi \in R$ let $\tilde{\delta}(\phi) \in \underline{e}(L)$ be given by

$$\tilde{\delta}(\phi) = \eta(\phi,\xi_\phi) \ . \qquad (4.2.1)$$

Theorem 4.2.1. *One has* $\tilde{\delta}(\phi) \in \underline{e}(L,\alpha)$ *and in fact*

$$\tilde{\delta}: R \longrightarrow \underline{e}(L,\alpha)$$

is a Lie algebra isomorphism. Furthermore one has a commutative diagram

$$\begin{array}{ccccccc} & & & \underline{e}(L,\alpha) & & & \\ & & \nearrow & & \searrow & & \\ 0 & \longrightarrow \underline{\underline{R}} & & \uparrow \tilde{\delta} & & \underline{\mathfrak{a}} & \longrightarrow 0 \\ & & \searrow & & \nearrow & & \\ & & & R & & & \end{array}$$

where the top and bottom exact sequence are those given in (3.3.1) *and* (4.1.4).

Proof. Since $d\phi = \beta_{\xi_\phi}$ one has $\tilde{\delta}(\phi) \in \underline{e}(L,\alpha)$ by Theorem 3.3.1. By Proposition 3.4.5 one has

$$[\tilde{\delta}(\phi_1), \tilde{\delta}(\phi_2)] = \eta\left(\xi_{\phi_1}\phi_2, \left[\xi_{\phi_1}, \xi_{\phi_2}\right]\right).$$

But since $\xi_{\phi_1}\phi_2 = [\phi_1,\phi_2]$ and since $\left[\xi_{\phi_1}, \xi_{\phi_2}\right] = \xi_{[\phi_1,\phi_2]}$ it follows that $\tilde{\delta}$ is a homomorphism. Obviously $\tilde{\delta}(\phi) = 0$ implies $\phi = 0$. But if $\eta = \eta(\phi,\xi) \in \underline{e}(L,\alpha)$ then ϕ is real and $\beta_\xi = d\phi$. Thus $\xi_\phi = \xi$, and hence $\eta = \tilde{\delta}(\phi)$. Thus $\tilde{\delta}$ is an isomorphism. That the diagram is commutative is obvious. QED

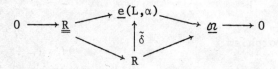

4.3. Because of the curvature ω it follows from Proposition 3.4.4 that S is not an $\underline{\mathfrak{a}}$-module using covariant differentiation. However a crucial point is that S becomes a module for the central

extension R of $\mathfrak{a}$. That is, the curvature may be "killed" using R.

Let

$$\delta: R \longrightarrow \text{End } S \qquad (4.3.1)$$

be defined by

$$\delta(\phi)\, s = \left(\nabla_{\xi_\phi} + 2\pi i\, \phi\right) s \qquad (4.3.2)$$

for any $\phi \in R$, $s \in S$.

Theorem 4.3.1. δ is a representation of R on S (i.e. $\delta[\phi,\psi] = [\delta(\phi), \delta\psi]$). Moreover for any $s \in S$ one has

$$\widetilde{\delta(\phi)\, s} = \tilde{\delta}(\phi)\, \tilde{s}\,. \qquad (4.3.3)$$

Proof. One has (4.3.3) by Proposition 3.4.2. But then δ is a representation since the map $S \longrightarrow \tilde{S}$ is an isomorphism and, by Theorem 4.2.1, $\tilde{\delta}$ defines a representation of R on $\tilde{S}$. QED

Remark 4.3.1. The map δ, which we call pre-quantization, is the first step in the "quantization" of functions (roughly speaking, the conversion of functions to operators). The second step involving polarization will be considered elsewhere. One recalls that during the early stages of the development of quantum mechanics a central mathematical question was finding a representation for the Heisenberg commutation relations. This of course was solved by the Stone - von Neumann theorem. The more general question involves an arbitrary finite dimensional Lie algebra $\mathfrak{g}$. The point of view here is that so that $\theta(\xi)\omega = 0$ (i.e. ξ is locally hamiltonian in a generalized sense - ω here is not necessarily non-singular) if and only if β_ξ is closed.

Proof. This follows immediately from (1.3.1) since $d\omega = 0$.

QED

Let $\underline{\mathfrak{n}}(X,\omega)$ (respectively $\underline{d}(X,\omega)$) be the Lie algebra of all (respectively real) vector fields ξ such that $\theta(\xi)\omega = 0$.

Remark 3.1.1. If $\xi \in \underline{d}(X)$ is globally integrable then clearly $\xi \in \underline{d}(X,\omega)$ if and only if the corresponding 1-parameter group is in $D(X,\omega)$.

One recalls that if $\beta \in \Omega(M)$ for some manifold M and $\xi \in \underline{d}(M)$ is globally integrable then

$$\big(\theta(\xi)\beta\big)(p) = \frac{d}{dt}\left[\big(\sigma(t)\cdot\beta\big)(p)\right]\Big|_{t=0}$$

where $\sigma(t)$ is the 1-parameter generated by ξ and the action of $D(M)$ on $\Omega(M)$ is defined by

$$\tau \cdot \beta = (\tau^*)^{-1}(\beta) . \tag{3.1.2}$$

3.2. Let $\underline{\mathfrak{a}}(X,\omega) \subseteq \underline{d}(X,\omega)$ be the subalgebra of all real vector fields ξ such that β_ξ is exact. That is $\beta_\xi = d\phi$ for some real $\phi \in C$.

Remark 3.2.1. Obviously $\underline{\mathfrak{a}} = \underline{d}(X,\omega)$ if $H^1(X, \underline{\underline{R}}) = 0$. In particular this equality holds if X is simply connected.

once a homomorphism of $\mathfrak{g}$ into R is given (e.g. the p_i, q_j and 1 for the Heisenberg Lie algebra) the representation is already naturally prescribed. It is given by δ together with the restriction imposed by a suitable polarization. A second question is whether the Lie algebra representation leads to a group representation. Under normal circumstances this is very difficult since the Hilbert space is abstract; one has only such tools as Nelson's theorem. However for the case at hand this is dealt with by (4.3.3). The relation (4.3.3) implies that the action of R on S is in effect by infinitesmal connection preserving transformations on L^*. By a result of Palais and Theorem 2.10.1 one thus is reduced to the question as to whether the hamiltonian vector fields ξ_ϕ for the appropriate $\phi \in R$ are globally integrable. Where this is the case, the action of the group on the Hilbert space rather than being abstract will arise from point-point transformations in L^* (see Theorem 4.5.1).

Even though in defining δ we are dealing with a specific line bundle with connection the representation δ depends, up to equivalence, only on the class ℓ of (L,α) as one easily sees.

Remark 4.3.2. Without defining polarization we can still point out the difference between pre-quantization and quantization. The point is that the choice of a polarization F of X has the effect of distinguishing both a Lie subalgebra R_F^1 of R and a subspace S_F of S. Moreover S_F is stable under $\delta(R_F^1)$ inducing a representation

$$\delta_F : R_F^1 \longrightarrow \text{End } S_F .$$

The map is what we call quantization.

It should be pointed out that both δ and δ_F depend on the choice of an element $\ell \in \mathcal{L}_c(X,\omega)$. (We of course assume $[\omega]$ is integral) and hence should be indexed as δ^{ℓ} and δ_F^{ℓ} . Fixing ℓ , $\mathcal{L}_c(X,\omega)$ is then isomorphic with $\Pi^* \cong H^1(X, \underline{T})$ so that we can regard $H^1(X, \underline{T})$ as parameterizing all possible prequantizations of R or quantizations of R_F^1 . As such this statement is a generalization of a theorem of Segal, see [3], p. 474. Indeed a special case of a polarized symplectic manifold is the cotangent bundle of a manifold M . Here $[\omega] = 0$ so that $[\omega]$ is integral. Also in this case S_F can be identified with all smooth functions on M and R_F^1 may be identified with all functions on X which at most are polynomials of the first degree on each cotangent space (linear in the p's). Essentially Segal defines quantization for R_F^1 in this case and observes that it is not unique but that $H^1(M, \underline{T})$ parameterizes all possible quantizations. But $H^1(M, \underline{T}) = H^1(X, \underline{T})$ and it may be shown that δ_F^{ℓ} over all $\ell \in \mathcal{L}_c(X,\omega)$ agrees with Segal's definition of quantization.

4.4. If a Lie group G with Lie algebra $\mathfrak{g}$ operates smoothly on a manifold M by a homomorphism $\sigma : G \longrightarrow D(M)$ we let

$$d\sigma : \mathfrak{g} \longrightarrow d(M)$$

be the homomorphism given by

$$\begin{aligned} \left(d\sigma(x)\phi\right)(p) &= \frac{d}{dt} \phi(\exp - tx \cdot p)\Big|_{t=0} \qquad (4.4.1) \\ &= \frac{d}{dt} \left((\exp tx \cdot \phi)(p)\right)\Big|_{t=0} \end{aligned}$$

where $x \in \mathfrak{g}$, $\phi \in D(M)$ and $p \in M$.

Remark 4.4.1. Note that if G operates on the manifolds X and Y by homomorphisms σ_X and σ_Y and $\tau: X \longrightarrow Y$ is a smooth map then τ is a G-map if and only if $d\sigma_X(x)$ and $d\sigma_Y(x)$ are τ-related for all $x \in \mathfrak{g}$. Indeed assume $x \in \mathfrak{g}$, $a = \exp x \in G$ and let $p \in X$. The trajectories of $\sigma_X(x)$ and $\sigma_Y(x)$ through p and τp respectively are the curves given by $\gamma_X(t) = \exp - tx \cdot p$ and $\gamma_Y(t) = \exp - tx \cdot p$. Now if $\gamma_X(x)$ and $\gamma_Y(y)$ are τ-related then $\tau(\gamma_X(t))$ is also a trajectory of $\sigma_Y(x)$ through τp. By the uniqueness of the trajectory one has $\tau\gamma_X(t) = \gamma_Y(t)$ for all t so that $\tau\, a \cdot p = a \cdot \tau\, p$. Since G is generated by $\exp \mathfrak{g}$ it follows that if $d\sigma_X(x)$ and $d\sigma_Y(x)$ are τ-related for all $x \in \mathfrak{g}$ then τ is a G-map. The other direction is obvious.

Assume that a connected Lie group G with Lie algebra $\mathfrak{g}$ operates smoothly on X by a homomorphism $\sigma: G \longrightarrow D(X)$. We will say that (X,ω) is a G-symplectic space X if ω is G-invariant (i.e. $\mathrm{Im}\sigma \subseteq D(X,\omega)$. Assume, until otherwise stated, that is the case and, for convenience, G is simply connected. Also, unless stated otherwise, we do not assume that G is transitive on X (i.e. X is not necessarily G-homogeneous).

We will say that (X,ω) is a G-strongly symplectic space or G is strongly symplectic on X if $\mathrm{Im}\sigma \subseteq \underline{\mathfrak{a}}$ so that one has a homomorphism

$$d\sigma: \mathfrak{g} \longrightarrow \underline{\mathfrak{a}} .$$

Remark 4.4.2. Note that if $[\omega]$ is integral then by the corrollary to Theorem 3.3.1 if $\ell \in \mathcal{L}_c(X,\omega)$ is arbitrary one has

$$\sigma \colon G \longrightarrow D_\ell(X) .$$

Proposition 4.4.1. *Strongly symplectic is equivalent to symplectic in either one of the following 2 cases:*

(1) X *is simply connected*

(2) $\mathfrak{g} = [\mathfrak{g},\mathfrak{g}]$, (e.g. *if* $\mathfrak{g}$ *is semi-simple*).

Proof. This follows immediately from Proposition 3.2.1 in the case of (2) and the equality $\underline{\mathfrak{a}} = \underline{d}(X,\omega)$ in the case of (1). QED

Remark 4.4.3. If X is only a G-symplectic space but not strongly symplectic, then for many purposes X may be replaced by its simply connected cover space X_s which has (since G is simply connected) a natural structure of a G-strongly symplectic homogeneous space in such a way that the diagram

$$\begin{array}{ccccc} G & \times & X_s & \longrightarrow & X_s \\ \downarrow & & & & \downarrow \\ G & \times & X & \longrightarrow & X \end{array}$$

is commutative.

Remark 4.4.4. Our definition of a strongly symplectic space is based on the Lie algebra. When dealing with disconnected groups one has to give a group theoretic rather than an infinitesmal definition of this notion. We shall not need this but it may be accomplished if $[\omega]$ is integral by introducing the following group $A = A(X,\omega) \subseteq D(X,\omega)$.

Any element $\gamma \in \Gamma$ defines a singular 1-cycle $\underline{\gamma}$. Let $D_o(X)$ be the subgroup of all diffeomorphisms $\tau \in D(X)$ which induce the trivial action on the first homology group. That is, $\tau \in D_o(X)$ if and only if for any $\gamma \in \Gamma$ the 1-cycle $\underline{\gamma} - \underline{\tau\gamma}$ is the boundary $\partial\rho$ of a piece-wise smooth singular 2-chain ρ . The integral $\int_\rho \omega$ is then well defined. One lets A be the subgroup of all $\tau \in D_o(X)$ such that

$$e^{2\pi i \int_\rho \omega} = 1$$

for any piece-wise smooth ρ such that $\partial\rho = \underline{\gamma} - \tau\underline{\gamma}$, for any $\gamma \in \Gamma$. It follows easily from Proposition 2.3.1 and the proof of (1.8.3) that A is the intersection of the $D_\ell(X,\omega)$ over all $\ell \in \mathcal{L}_c(X,\omega)$. Furthermore, (see Remark 3.3.2) one has $\operatorname{Im}\sigma \subseteq A$ if and only if $\operatorname{Im} d\sigma \subseteq \underline{\mathfrak{a}}$ so that strongly symplectic can be defined group theoretically.

4.5. If $[\omega]$ is integral so that $\ell = [(L,\alpha)] \in \mathcal{L}_c(X,\omega)$ exists we introducedin § 1.13 the notion of a lifting $\nu : G \longrightarrow E(L,\alpha)$ of σ . We say the lifting is smooth if G operates smoothly on L^* .

Here without assuming $[\omega]$ is integral but that G is strongly symplectic on X we say that a homomorphism

$$\lambda : \mathfrak{g} \longrightarrow R$$

is a lifting of $d\sigma$ in case the diagram

$$\begin{array}{ccccccccc} 0 & \longrightarrow & \underline{\underline{R}} & \longrightarrow & R & \longrightarrow & \underline{\mathfrak{a}} & \longrightarrow & 0 \\ & & & & & \nwarrow^{\lambda} & \uparrow{\scriptstyle d\sigma} & & \\ & & & & & & \mathfrak{g} & & \end{array} \qquad (4.5.1)$$

is commutative.

Remark 4.5.1. Note that λ exists if and only if the class $[\mu] \in H^2(\mathfrak{g}, \underline{\underline{R}})$ vanishes where μ is the cocycle given by $\mu(x,y) = ([\mu_o(x), \mu_o(y)] - \mu_o[x,y]) \in R$ for $\mu_o : \mathfrak{g} \longrightarrow R$ any linear map making (4.5.1) commutative. In particular λ always exists if $H^2(\mathfrak{g}, \underline{\underline{R}}) = 0$ (e.g. if $\mathfrak{g}$ is semi-simple).

Note also that if λ is a lifting then so is $\lambda + f$ where $f \in \mathfrak{g}'$ (the real dual to $\mathfrak{g}$) is such that $f|[\mathfrak{g},\mathfrak{g}] = 0$, (i.e. $f: \mathfrak{g} \longrightarrow \underline{\underline{R}}$ is a Lie algebra homomorphism, or equivalently $f \in H^1(\mathfrak{g}, \underline{\underline{R}})$). Moreover any lifting is uniquely of the form $\lambda + f$.

The two notions of lifting correspond.

Theorem 4.5.1. *Assume* $[\omega]$ *is integral and let* $\ell = [(L,\alpha)] \in \mathcal{L}_c(X,\omega)$. *Also assume that* (X,ω) *is a* G-*symplectic space. Then* $\sigma = \sigma_X$ *may be lifted (which implies* $\operatorname{Im}\sigma \subseteq D_\ell(X)$) *smoothly to* $\sigma_L : G \longrightarrow E(L,\alpha)$, *so that one has a commutative diagram,*

$$\begin{array}{ccccccc} 0 & \longrightarrow & \underline{\underline{T}} & \longrightarrow & E(L,\alpha) & \longrightarrow & D_\ell(X) & \longrightarrow & 1 \\ & & & & & \nwarrow^{\sigma_L} & \uparrow \sigma_X & & \\ & & & & & & G & & \end{array} \qquad (4.5.2)$$

if and only if G *is strongly symplectic on* X *and* $d\sigma$ *may be lifted to* $\lambda : \mathfrak{g} \longrightarrow \underline{\underline{R}}$. *Moreover in such a case* σ_L *may be uniquely chosen so that it is related to* λ *by the equality of maps* $\mathfrak{g} \longrightarrow \underline{e}(L,\alpha)$

$$d\sigma_L = \tilde{\delta} \circ \lambda \qquad (4.5.3)$$

(*see Theorem* 4.2.1).

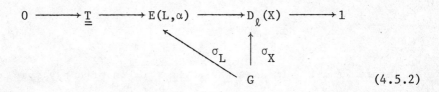

Proof. Assume G is strongly symplectic and let λ be a lifting of $d\sigma$. Then, by Theorem 4.2.1, $\tilde{\delta} \circ \lambda\colon \mathfrak{g} \longrightarrow \underline{e}(L, \alpha)$ is a homomorphism such that $\pi_*(\tilde{\delta} \circ \lambda(x)) = d\sigma(x)$ for any $x \in \mathfrak{g}$. But now since $d\sigma(x)$ is globally integrable the same is true for $(\tilde{\delta} \circ \lambda)(x)$ by Theorem 2.10.1. But now by a theorem of Palais (see Theorem III, page 91, [3]) there exists a unique smooth action of G on L^* defining a homomorphism $\sigma_L\colon G \longrightarrow D(L^*)$ such that $d\sigma_L = \tilde{\delta} \circ \lambda$. However one clearly has $\operatorname{Im}\sigma_L \subseteq E(L, \alpha)$ (see Remark 3.3.1). To show that σ_L is a lifting of σ we have that $\tilde{\pi}$ is a G-map but this is clear from Remark 4.4.1 since $d\sigma_L(x) = (\tilde{\delta} \circ \lambda)(x)$ is $\tilde{\pi}$-related to $d\sigma(x)$ for all $x \in \mathfrak{g}$.

Now assume σ_L is a lifting of σ. We have only to show that G is strongly symplectic on X and there exists a lifting λ of $d\sigma$ such that (4.5.3) is satisfied. But now $d\sigma_L\colon \mathfrak{g} \longrightarrow \underline{e}(L, \alpha)$ is a homomorphism such that

$$\pi_* \circ d\sigma_L = d\sigma. \tag{4.5.4}$$

But, by Theorem 3.3.1, $\operatorname{Im} d\sigma \subseteq \underline{\mathfrak{a}}$ so that G is strongly symplectic on X. But also, by Theorem 4.2.1, $d\sigma_L$ is uniquely of the form $d\sigma_L = \tilde{\delta} \circ \lambda$ where $\lambda\colon \mathfrak{g} \longrightarrow R$ is a homomorphism. But finally (4.5.4) also asserts that λ is a lifting of $d\sigma\colon \mathfrak{g} \longrightarrow \underline{\mathfrak{a}}$. QED

5. Orbits and Hamiltonian G-spaces. We now regard a simply connected Lie group G with Lie algebra $\mathfrak{g}$ to be fixed.

A symplectic manifold (X, ω) will be called a Hamiltonian G-space if one is given a homomorphism

$$\lambda\colon \mathfrak{g} \longrightarrow R$$

(R, recall is the space of all smooth real functions on X regarded as a Lie algebra under Poisson bracket) such that

(1) $(d\lambda(x))p$ over all $x \in \mathfrak{g}$ spans the dual space to the tangent space at p for all $p \in X$

(2) $\xi_{\lambda(x)}$ is globally integrable for all $x \in \mathfrak{g}$.

Remark 5.1.1. Note that by Palais' theorem this implies that (X, ω) is a G-strongly symplectic space where (since $\xi_{\lambda(x)} \in \mathfrak{a}$)

$$d\sigma(x) = \xi_{\lambda(x)} \tag{5.1.1}$$

so that one has a commutative diagram.

$$\begin{array}{ccccccccc} 0 & \longrightarrow & \mathbb{R} & \longrightarrow & R(X) & \longrightarrow & \_(X) & \longrightarrow & 0 \\ & & & & & \nwarrow^{\lambda} & \uparrow d\sigma & & \\ & & & & & & \mathfrak{g} & & \end{array}$$

and (1) implies that X is a G-homogeneous space (since $(\xi_{\lambda(x)})_p$, for any $p \in X$, over all $x \in \mathfrak{g}$ must span the tangent space at p). Conversely if (X, ω) is a homogeneous G-strongly symplectic space such that $d\sigma$ can be lifted then (X, ω) together with a <u>lifting</u> λ gives (X, ω) the structure of a Hamiltonian G-space.

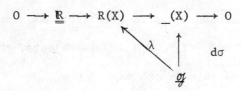

A Hamiltonian G-space is denoted by the triple (X, ω, λ) where we will generally write ω_X and λ_X for ω and λ to be specific. Also the action of G on X is denoted by the homomorphism $\sigma_X\colon G \longrightarrow D(X, \omega_X)$. In the light of Theorem 4.5.1 the significance of the notion of Hamiltonian G-space is that whenever a line bundle with connection (L, α) is given over X where $\omega_X = \text{curv}(L, \alpha)$ there is a distinguished lifting $\sigma_L: G \longrightarrow E(L, \alpha)$ of σ_X. That is one has the following corollary (or really a reformulation of a part of) of Theorem 4.5.1.

Theorem 5.1.1. Let (X, ω_X, λ_X) be a Hamiltonian G-space. Assume $[\omega_X] \in H^2(X, \mathbb{R})$ is integral and let $[(L, \alpha)] \in \mathcal{L}_c(X, \omega_X)$. Then there is a unique lifting

$$\sigma_L: G \longrightarrow E(L, \alpha) \tag{5.1.2}$$

of σ_X such that $d\sigma_L = \tilde{\delta} \circ \lambda_X$.

Proof. Immediate from Theorem 4.5.1. QED

We may regard the Hamiltonian G-spaces as objects of a category Ham(G) where if (X, ω_X, λ_X) and (Y, ω_Y, λ_Y) are objects, then a map of the first into the second is a smooth map

$$\tau: X \longrightarrow Y \tag{5.1.3}$$

of manifolds such that

$$\tau^* \omega_Y = \omega_X \quad \text{and} \quad \lambda_Y(x) \circ \tau = \lambda_X(x) \tag{5.1.4}$$

for all $x \in \mathfrak{g}$.

Proposition 5.1.1. The map (5.1.3) is a G-map so that $\xi_{\lambda_X}(x)$ and $\xi_{\lambda_Y}(x)$ are τ-related for any $x \in \mathfrak{g}$. Moreover for any $p \in X$

$$\tau_*: T_p(X) \longrightarrow T_{\tau p}(Y) \tag{5.1.5}$$

<u>is an isomorphism and in fact</u> τ <u>is a covering map of manifolds</u>.

<u>Proof</u>. Since $\tau \circ \lambda_Y(x) = \lambda_X(x)$ one has $\tau^* d(\lambda_Y(x)) = d\lambda_X(x)$ or $\tau^*\left(i(\xi_{\lambda_Y}(x))\omega_Y\right) = i(\xi_{\lambda_X(x)})\omega_X = i(\xi_{\lambda_X(x)})\ \tau^*\omega_Y$. But then $\tau_*\left((\xi_{\lambda_X(x)})\ p\right) = (\xi_{\lambda_Y(x)})_{\tau p}$ for any $p \in X$ since ω_Y is non-singular. Thus $\xi_{\lambda_X(x)}$ and $\xi_{\lambda_Y(x)}$ are τ-related. Thus τ is G-map by Remark 4.4.1. This also shows that the map (5.1.5) is surjective since $T_p(X)$ is spanned by $(\xi_{\lambda_X(x)})_p$ over all $x \in \mathfrak{g}$. But it is injective since $\tau^*\omega_Y = \omega_X$ and ω_X is non-singular. This proves (5.1.5) is an isomorphism and also that τ is a covering map. Indeed if $p \in X$ and G_p, $G_{\tau p}$ are the isotropy subgroups of G at p and τp, respectively, so that as homogeneous spaces

$$X \cong G/G_p , \qquad Y \cong G/G_{\tau p}$$

one has

$$G_o \subseteq G_p \subseteq G_{\tau p}$$

where G_o is the identity component of $G_{\tau p}$. QED

If $X = (X,\omega_X,\lambda_X)$ and $Y = (Y,\omega_Y,\lambda_Y)$ are Hamiltonian G-spaces and a map $\tau: X \longrightarrow Y$ of such spaces exist we will say simply that X covers Y and τ is a covering map of Hamiltonian G-spaces. Of course τ is an isomorphism in Ham(G) if and only if it is a diffeomorphism of manifolds.

Remark 5.1.2 Note that if

$$\tau: X \longrightarrow Y$$

is a covering of manifolds where (Y,ω_Y,λ_Y) is a Hamiltonian G-space, then there exists a unique structure (X,ω_X,λ_X) of a Hamiltonian G-space on X such that τ is a map of Hamiltonian G-spaces. Indeed one defines ω_X and λ_X by (5.1.3). $\xi_{\lambda_X(x)}$ is globally integrable since the lifting of a global trajectory of $\xi_{\lambda_Y(x)}$ is a global trajectory of $\xi_{\lambda_X(x)}$.

5.2. We now give examples of Hamiltonian G-spaces. If $a \in G$ and $x \in \mathfrak{g}$ we write $a \cdot x$ for $\mathrm{Ad}\, a\, x$. The real dual $\mathfrak{g}'$ of $\mathfrak{g}$ is a G-module by the contragradient Ad' of the adjoint representation. Writing $a \cdot f$ for $(\mathrm{Ad}'a)\, f$ when $a \in G$ and $f \in \mathfrak{g}'$, one has

$$\langle a \cdot f,\ x \rangle = \langle f,\ a^{-1} \cdot x \rangle .$$

But then $\mathfrak{g}'$ is a $\mathfrak{g}$-module where $y \cdot f = \frac{d}{dt}(\exp ty \cdot f)\Big|_{t=0}$ for any $y \in \mathfrak{g}$ and one has the relation

$$\langle y \cdot f\ ,\ x \rangle = \langle f,\ [x,y] \rangle . \tag{5.2.1}$$

A g-orbit in $\mathfrak{g}'$ is any set $0 \subseteq \mathfrak{g}'$ of the form $0 = G \cdot f$ where $f \in \mathfrak{g}'$. An orbit has the structure of a G-homogeneous space and in fact one has the equivalence

$$0 \cong G/G_f$$

of such spaces where $0 = G \cdot f$ and $G_f = \{a \in G | a \cdot f = f\}$. We will show that any orbit has a natural structure of a Hamiltonian G-space.

First of all $\mathfrak{g}'$ is a manifold on which G operates smoothly and hence one has a homomorphism $\mathfrak{g} \longrightarrow \underline{d}(\mathfrak{g}')$, $x \longrightarrow \eta^x$ where η^x is the vector field on $\mathfrak{g}'$ given by

$$(\eta^x \phi)(f) = \frac{d}{dt} \phi (\exp -tx \cdot f) \Big|_{t=0} \tag{5.2.2}$$

for any $\phi \in C(\mathfrak{g}')$, $f \in \mathfrak{g}'$.

Now for any $y \in \mathfrak{g}$ let $\psi^y \in C(\mathfrak{g}')$ be the (smooth) function on $\mathfrak{g}'$ given by

$$\psi^y(f) = \langle f,y \rangle . \tag{5.2.3}$$

Remark 5.2.1. Obviously $d\psi^y$, over all $y \in \mathfrak{g}$, span the cotangent space to $\mathfrak{g}'$ at any point of $\mathfrak{g}'$.

Lemma 5.2.1. <u>For any</u> $x, y \in \mathfrak{g}$ <u>one has</u>

$$\eta^x \psi^y = \psi^{[x,y]} \tag{5.2.4}$$

<u>so that for any</u> $f \in \mathfrak{g}'$ <u>one has</u>

$$\begin{aligned} (\eta^x \psi^y)(f) &= \langle f, [x,y] \rangle \\ &= -\langle x \cdot f, y \rangle . \end{aligned} \tag{5.2.5}$$

<u>Proof</u>. By definition

$$\begin{aligned} (\eta^x \psi^y)(f) &= \frac{d}{dt} \psi^y(\exp -t x \cdot f) \Big|_{t=0} \\ &= \frac{d}{dt} \langle \exp -tx \cdot f,y\rangle \Big|_{t=0} . \end{aligned}$$

But the right side equals $\langle -x \cdot f, y \rangle = \langle f, [x,y] \rangle = \psi^{[x,y]}(f)$.
Hence $\eta^{x} \psi^{y} = \psi^{[x,y]}$. QED

Now for any $f \in \mathfrak{g}'$ let $\mathfrak{g}_f = \{x \in \mathfrak{g} \mid x \cdot f = 0\}$ so that $\mathfrak{g}_f$ is the Lie algebra of G_f . Since clearly $(\eta^x)_f \in T_f(0)$ where $0 = G \cdot f$, let

$$\sigma_f : \mathfrak{g} \longrightarrow T_f(0)$$

be the linear map given by putting $\sigma_f(x) = (\eta^x)_f$.

Proposition 5.2.1. <u>One has an exact sequence</u>

$$0 \longrightarrow \mathfrak{g}_f \xrightarrow{\text{inj}} \mathfrak{g} \xrightarrow{\sigma_f} T_f(0) \longrightarrow 0 \quad .$$

<u>Proof</u>. σ_f is surjective since $0 = G \cdot f$ is G-homogeneous. On the other hand the kernel of σ_f is $\mathfrak{g}_f$ by Remark 5.2.1 and (5.2.5). QED

The symplectic structure on any orbit $0 \subseteq \mathfrak{g}'$ will arise from

Proposition (Kirillov) 5.2.2. <u>Let</u> $0 \subseteq \mathfrak{g}'$ <u>be any</u> G-<u>orbit and let</u> $f \in 0$. <u>Then there exists a unique alternating</u> 2-<u>form</u> ω_f <u>in</u> $T_p(0)$ <u>such that for any</u> $x, y \in \mathfrak{g}$ <u>one has</u>

$$\omega_f(\sigma_f y, \sigma_f x) = \langle f, [x,y] \rangle \; .$$

<u>Moreover</u> ω_f <u>is non-singular so that</u> 0 <u>is even-dimensional</u>.

<u>Proof</u>. Let B_f be the alternating 2-form on $\mathfrak{g}$ defined by putting $B_f(y,x) = \langle f, [x,y] \rangle = \langle -x \cdot f, y \rangle$. Thus B_f induces

a non-singular alternating form on $\mathfrak{g}/\mathfrak{g}_{(f)}$ where $\mathfrak{g}_f = \{x \in \mathfrak{g} \mid B_f(y,x) = 0$ for all $y \in \mathfrak{g}\}$. But from the relation $B_f(y,x) = \langle - x\cdot f, y\rangle$ one has $\mathfrak{g}_f = \mathfrak{g}_{(f)}$. However the map σ_f induces an isomorphism $\mathfrak{g}/\mathfrak{g}_f \cong T_f(0)$ by Proposition 5.2.1. This proves the Proposition 5.2.2.

QED

5.3. Now fix an orbit $0 \subseteq \mathfrak{g}'$. The action of G on 0 induces a homomorphism of groups $\sigma_0 \colon G \longrightarrow D(0)$ and a homomorphism of Lie algebras

$$d\sigma_0 : \mathfrak{g} \longrightarrow \underline{d}(0)$$

where if $\xi^x = d\sigma_0(x)$ one has, by (5.2.2)

$$\xi^x = \eta^x \Big| 0 \tag{5.3.1}$$

for any $x \in \mathfrak{g}$.

Next for any $x \in \mathfrak{g}$ let $\phi^x = \psi^x \Big| 0$ so that $\phi^x \in C(0)$.

By Lemma 5.2.1. one has

$$\xi^x \phi^y = \phi^{[x,y]} . \tag{5.3.2}$$

Finally let ω_0 be the 2-form (not yet shown to be smooth or even continuous) on 0 defined so that $(\omega_0)_f = \omega_f$.

Proposition 5.3.1. ω_0 <u>is smooth and for any</u> $y \in \mathfrak{g}$ <u>one has</u>

$$i(\xi^y)\omega_0 = d\phi^y .$$

<u>Proof</u>. If $f \in 0$ then $\sigma_f y = (\xi^y)_f$ by (5.3.1). But then

by Proposition 5.2.2. one has, for $x, y \in \mathfrak{g}$

$$\omega_0(\xi^y, \xi^x) = \phi^{[x,y]} \tag{5.3.3}$$

since $\langle f, [x,y] \rangle = \psi^{[x,y]}(f) = \phi^{[x,y]}(f)$ for any $f \in 0$.

But since 0 is homogeneous there exists for any $f \in 0$ elements $x_i \in \mathfrak{g}$, $i = 1,\ldots,k$ such that $(\xi^{x_i})_g$ are a basis of $T_g(0)$ for all g in some neighborhood of f . Since $\omega_0(\xi^{x_i}, \xi^{x_j}) = \phi^{[x_i, x_j]}$ is smooth in this neighborhood it follows that ω_0 is smooth on 0 . Fixing $y \in \mathfrak{g}$ the relations (5.3.2) and (5.3.3) imply that

$$\left\langle (i\ (\xi^y)\omega_0)\ ,\ \xi^x \right\rangle = \omega_0(\xi^y, \xi^x) = \xi^x\phi^y = \left\langle d\phi^y, \xi^x \right\rangle$$

for all $x \in \mathfrak{g}$. Since the ξ^x span the tangent space at each point of 0 the proposition is proved. QED

Now where $R(0)$ is the set of all real smooth functions on 0 let

$$\lambda_0 : \mathfrak{g} \longrightarrow R(0)$$

be the linear map given by $\lambda_0(x) = \phi^x$ so that if $f \in 0$

$$\lambda_0(x)(f) = \langle f,x \rangle .$$

Not only is $(0,\omega_0)$ a symplectic manifold but in fact one has

Theorem 5.3.1. <u>The pair $(0, \omega_0)$ is a G-strongly symplectic manifold for any orbit $0 \subseteq \mathfrak{g}'$. Moreover λ_0 is a lifting of $d\sigma_0$ so that λ_0 is a Lie algebra homomorphism ($R(0)$ is a Lie algebra</u>

under Poisson bracket) and $(0, \omega_0, \lambda_0)$ is a Hamiltonian G-space.

Proof. We first show that ω_0 is G-invariant or $\theta(\xi^x)\omega_0 = 0$ for all $x \in \mathfrak{g}$.

Apply $\theta(\xi^x)$ to both sides of (5.3.3). One has $\theta(\xi^x)\, d\phi^y = d\xi^x\, \phi^y = d\phi^{[x,y]}$ by (5.3.2.). But $d\phi^{[x,y]} = i(\xi^{[x,y]})\omega_0$ by Proposition 5.3.1. so that

$$\theta(\xi^x)\, d\phi^y = i\,(\xi^{[x,y]})\,\omega_0 \quad . \tag{5.3.4}$$

On the other hand $[\theta(\xi^x), i\,(\xi^y)] = i\,([\xi^x, \xi^y])$ by (1.3.2). But this equals $i(\xi^{[x,y]})$ since σ_0 is a homomorphism. Thus $\theta(\xi^x)\, i(\xi^y)\,\omega_0 = i\,(\xi^{[x,y]})\,\omega_0 + i(\xi^y)\,\theta(\xi^x)\,\omega_0$. Comparing with (5.3.4) and Proposition 5.3.1 one has

$$i(\xi^y)\,\theta(\xi^x)\,\omega_0 = 0$$

for all $x,y \in \mathfrak{g}$. But since the ξ^y over all y span the tangent space at any point of 0 one has $\theta(\xi^x)\,\omega_0 = 0$ for all $x \in \mathfrak{g}$ and hence ω_0 is G-invariant.

But now (see (1.3.1)) $\theta(\xi^x) = i(\xi^x)\, d + d\, i(\xi^x)$ and hence the relation $\theta(\xi^x)\,\omega_0 = 0$ implies that $i(\xi^x)\, d\omega_0 + d\, i(\xi^x)\,\omega_0 = 0$. But $i(\xi^x)\,\omega_0 = d\phi^x$ by Proposition 5.3.1 and hence $d\left(i(\xi^x)\,\omega_0\right) = d\, d\phi^x = 0$. Thus

$$i\,(\xi^x)\, d\omega_0 = 0$$

for all $x \in \mathfrak{g}$. Hence $d\omega_0 = 0$ since the ξ^x, over all $x \in \mathfrak{g}$, span the tangent space at 0 at any point. Thus $(0, \omega_0)$ is a symplectic manifold. But again by Proposition 5.3.1, $\beta_{d\sigma_0(x)} = i(d\sigma_0(x))\omega_0 = i(\xi^x)\omega_0 = d\phi^x$ is exact and hence $\mathrm{Im}\, d\sigma_0 \subseteq \underline{\mathfrak{a}}$ so that G is strongly symplectic on X. But by definition of ξ_ϕ this implies that for any $x \in \mathfrak{g}$

$$\xi_{\phi^x} = \xi^x = d\sigma_0(x) \quad . \tag{5.3.5}$$

Thus to show that $\lambda_0(x) = \phi^x$ defines a lifting of $d\sigma_0$ we have only to show that λ_0 is a homomorphism. That is

$$[\phi^x, \phi^y] = \phi^{[x,y]} \quad .$$

But by definition $[\phi^x, \phi^y] = \xi_{\phi^x}\, \phi^y = \xi^x\, \phi^y$ by (5.3.5). But $\xi^x\, \phi^y = \phi^{[x,y]}$ by (5.3.2). By Remark 5.2.1 the forms $d\phi^x$ over all $x \in \mathfrak{g}$ span the cotangent space any point of 0. Thus $(0, \omega_0, \lambda_0)$ is a Hamiltonian G-space. QED

According to Remark 5.1.2. it now follows that any manifold which covers a G-orbit in $\mathfrak{g}'$ has the structure of a Hamiltonian G-space. We now show conversely that the set of G-orbits in $\mathfrak{g}'$ are universal in the sense that any Hamiltonian G-space $X = (X, \omega_X, \lambda_X)$ covers an orbit 0 in $\mathfrak{g}'$ not only as a manifold but as a Hamiltonian G-space. Moreover when taken in the latter sense the orbit 0 is unique and the covering map is unique.

5.4. If X is simply connected as a manifold we will say that the Hamiltonian G-space X is simply connected.

Theorem 5.4.1. <u>Let</u> (X, ω_X, λ_X) <u>be any Hamiltonian</u> G-<u>space</u>. <u>Then there exists a unique orbit</u> 0 <u>such that a map</u>

$$\tau_X : X \longrightarrow 0 \tag{5.4.1}$$

<u>of Hamiltonian</u> G-<u>spaces exists</u>. <u>Moreover</u> τ_X <u>is unique</u>. <u>Furthermore</u> (5.4.1) <u>sets up a bijection between the set of all isomorphism classes of simply connected Hamiltonian</u> G-<u>spaces and all</u> G-<u>orbits</u> 0 <u>in</u> $\mathfrak{g}'$.

<u>Proof</u>. Let

$$\tau_X : X \longrightarrow \mathfrak{g}'$$

be the map given by

$$\langle \tau_X(p), x \rangle = \big(\lambda_X(x)\big)(p) \tag{5.4.2}$$

for any $x \in \mathfrak{g}$. Since $\psi^x\big(\tau_X(p)\big) = \langle \tau_X(p), x \rangle$ one has

$$\psi^x \circ \tau_X = \lambda_X(x) \quad . \tag{5.4.3}$$

Since the functions ψ^x, $x \in \mathfrak{g}$, define a coordinate system on $\mathfrak{g}'$ and since $\lambda_X(x) \in C(X)$ it follows from (5.4.3) that τ_X is smooth.

Now if $p \in X$ and $y \in \mathfrak{g}$ let

$$v = (\tau_X)_* \big(d\sigma_X(y)\big)_p \in T_{\tau_X(p)}(\mathfrak{g}') \; .$$

Then by (5.4.3)

$$\begin{aligned} v\psi^x &= \big(d\sigma_X(y)\big)_p \, \lambda_X(x) \\ &= (\lambda_X[y,x])(p) \end{aligned}$$

since $d\sigma_X(y) = \xi_{\lambda_X(y)}$ so that $d\sigma_X(y)\,\lambda_X(x) = [\lambda_X(y), \lambda_X(x)] = \lambda_X[y,x]$. But $(\lambda_X\,[y,x])(p) = \psi^{[y,x]}\big(\tau_X(p)\big)$ by (5.4.3). Thus

$$v\,\psi^x = \psi^{[y,x]}\big(\tau_X(p)\big)$$

for all $x \in \mathfrak{g}$. But also $(\eta^y)_{\tau_X(p)}\,\psi^x = \psi^{[y,x]}\big(\tau_X(p)\big)$ for all $x \in \mathfrak{g}$ by (5.2.4) so that $v = (\eta^y)_{\tau_X(p)}$. But then by (5.2.4) $\sigma_X(y)$ and η^y are τ_X-related and that

$$(\tau_X)_*\,\sigma_x(y) = \eta^y \quad .$$

But then τ_X is G-map since it is a $\mathfrak{g}$-map (see Remark 4.4.1). But also since G is transitive on X it follows that the image of τ_X is a G-orbit $0 \subseteq \mathfrak{g}'$. But then $\sigma_X(x)$ and ξ^x are τ_X-related for any $x \in \mathfrak{g}$ so that if $x,y \in \mathfrak{g}$ one has

$$\omega_0(\xi^y, \xi^x) \circ \tau_X = \big((\tau_X{}^*)\,\omega_0\big)\big(\sigma_X(y), \sigma_X(x)\big) \quad . \tag{5.4.4}$$

Now (5.4.3) implies

$$\phi^x \circ \tau_X = \lambda_X\,(x) \quad . \tag{5.4.5}$$

But $\omega_0(\xi^y, \xi^x) = \phi^{[x,y]}$ by (5.3.3) and hence the left side of (5.4.4) equals $\lambda_X([x,y])$. However $\omega_X\big(\sigma_X\,(y), \sigma_X\,(x)\big) = \lambda_X\,([x,y])$ by (4.1.6). But then $\omega_X = \tau_X{}^*\,\omega_0$ since the vector fields $\sigma_X(x)$, $x \in \mathfrak{g}$, span the tangent space at any point of X . But then by (5.4.5) τ_X is the map of Hamiltonian G-spaces. Now if 0_1 is any orbit and $\tau\colon X \longrightarrow 0_1$ is a map of Hamiltonian G-spaces, then one has $\big(\lambda_X(x)\big)(p) = \big(\lambda_{0_1}(x)\big)\big(\tau(p)\big)$ for any $p \in X$ and $x \in \mathfrak{g}$. But

$\lambda_{0_1}(x)\left(\tau(p)\right) = \langle \tau(p), x \rangle$. But $\left(\lambda_X(x)\right)(p) = \langle \tau_X(p), x \rangle$.

Thus $\tau_X = \tau$ which implies also that $0 = 0_1$. This proves the uniqueness of 0 and τ_X .

Now where X and Y are simply connected Hamiltonian G-spaces assume $\tau_X(X) = \tau_Y(Y) = 0 \subseteq \mathfrak{g}'$. Since X and Y are coverings of 0 as manifolds, if $\tau_X(p) = \tau_Y(q)$ for $p \in X$ and $q \in Y$, there exists a unique diffeomorphism $\tau\colon X \longrightarrow Y$ such that $\tau_Y \circ \tau = \tau_X$. But then $\omega_X = \tau_X^* \omega_0 = \tau^* \tau_Y^* \omega_0 = \tau^* \omega_Y$. Also if $x \in \mathfrak{g}$ then

$\lambda_X(x) = \phi^x \circ \tau_X = (\phi^x \circ \tau_Y) \circ \tau = \lambda_Y(x) \circ \tau$. Thus τ is an isomorphism of Hamiltonian spaces so that X and Y are isomorphic.

On the other hand if $\tau_X(X) \neq \tau_Y(Y)$ then X and Y cannot be isomorphic since if $\tau\colon X \longrightarrow Y$ is an isomorphism of Hamiltonian G-spaces then $\tau_Y \circ \tau$ and τ_X are distinct coverings by X of an orbit. This contradicts what has been proved. Thus the correspondence $X \longrightarrow \tau_X X$ sets up an injective map from the set of all isomorphism classes of simply connected Hamiltonian G-spaces into the set of all G-orbits in $\mathfrak{g}'$. But the correspondence is clearly surjective since by Remark 5.1.2 the simply connected covering space X of an orbit 0 has a structure of a Hamiltonian G-space such that the covering map τ is a map of Hamiltonian G-spaces (so that $\tau = \tau_X$). QED

Remark 5.4.1. Note that by Theorem 5.4.1 distinct orbits in $\mathfrak{g}'$ define non-isomorphic Hamiltonian G-spaces since τ_0 is clearly the identity map for any G-orbit $0 \subseteq \mathfrak{g}'$.

Corollary to Theorem 5.4.1. *Let* $X = (X, \omega_X, \lambda_X)$ *be a Hamiltonian G-space. Then* X *is isomorphic to an orbit* 0 *if and only if the functions* $\lambda_X(x)$, $x \in \mathfrak{g}$, *on* X, *separate points. That is if and only if for all* $p,q \in X$ *where* $p \neq q$ *there exists* $x \in \mathfrak{g}$ *such that* $(\lambda_X(x))(p) \neq (\lambda_X(x))(q)$.

Proof. Let $0 \subseteq \mathfrak{g}'$ be a G-orbit. Since $\lambda_0(x)(f) = \langle f,x \rangle$ for $f \in 0$, $x \in \mathfrak{g}$, it is obvious that the $\lambda_0(x)$, over all x, separate points. Hence if X and 0 are isomorphic the functions $\lambda_X(x)$, $x \in \mathfrak{g}$, separate points. Conversely if these functions separate points then the map τ_X which is explicitly given by (5.4.2) is clearly injective. Since in any case τ_X is a local diffeomorphism and a surjective map onto the G-orbit $\tau_X(X)$, it follows that τ_X is a diffeomorphism and hence an isomorphism of Hamiltonian G-spaces.

QED

5.5. If one drops the lifting λ_X and considers only the G-symplectic space (X, ω_X) then it is possible that if an orbit covering exists (i.e. a G-map $\tau: X \longrightarrow 0$ which is a covering of manifolds such that $\tau^* \omega_0 = \omega_X$) then τ or 0 need not be unique. For example if $\hat{\mathfrak{g}} = H^1(X, \mathbb{R})$ is the set of all $f \in \mathfrak{g}'$ such that $f|[\mathfrak{g},\mathfrak{g}] = 0$ it is easy to see that the elements of $\hat{\mathfrak{g}}$ are fixed points under the action of G on $\hat{\mathfrak{g}}'$ and conversely every fixed point (i.e., an orbit consisting of one point) is an element of $\hat{\mathfrak{g}}$. (This is clear since $x \cdot f = 0$ for all $x \in \mathfrak{g}$ if and only if $f|[\mathfrak{g},\mathfrak{g}] = 0$). Then if X is the trivial G-symplectic space, i.e. consisting of one point it covers every element of $\hat{\mathfrak{g}}$.

We now find that more generally the elements of $\hat{\mathfrak{g}}$ describe and limit the ambiguity for all possible orbit (in $\mathfrak{g}'$) coverings.

Theorem 5.5.1. *Assume* (X, ω_X) *is a G-symplectic homogeneous space and* $\tau: X \longrightarrow O \subseteq \mathfrak{g}'$ *is an orbit covering (in the sense of G-symplectic homogeneous spaces). Then if* $f \in \hat{\mathfrak{g}}$ *the set* $f + O = O^f$ *is again an orbit in* $\mathfrak{g}'$ *and the map* $\tau^f : X \longrightarrow O^f$ *given by*

$$\tau^f(p) = f + \tau(p)$$

is an orbit covering.

Conversely every orbit covering $\tau': X \longrightarrow O'$ *is uniquely of this form so that in particular* $O' = g + O$ *for some* $g \in \hat{\mathfrak{g}}$.

Proof. Assume that $\tau_i: X \longrightarrow O_i$, $1 = 1,2$ are two orbit coverings. Let $\lambda^i_X: \mathfrak{g} \longrightarrow R(X)$, $i = 1,2,$ be defined by putting $\lambda^i_X(x) = \psi^x \circ \tau_i$. Then (see Remark 5.1.2) λ^i_X , $i = 1,2$ are liftings of $d\sigma_X$

$$\begin{array}{ccccccccc} 0 & \longrightarrow & \underline{\underline{R}} & \longrightarrow & R(X) & \longrightarrow & \mathfrak{a}(X) & \longrightarrow & 0 \\ & & & & & \nwarrow {\scriptstyle \lambda^i_X} & \uparrow {\scriptstyle d\sigma_X} & & \\ & & & & & & \mathfrak{g} & & \end{array}$$

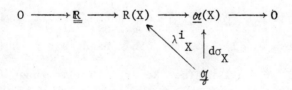

But then from the exact sequence above $\lambda^2_X(x) - \lambda^1_X(x) = \langle g,x \rangle \in \underline{\underline{R}}$ for all $x \in \mathfrak{g}$ for some unique $g \in \mathfrak{g}'$. But $g \in \hat{\mathfrak{g}}$ since the extension above is central. Since $\langle \tau^i(p) , x \rangle = \lambda^i_X(x)(p)$ this implies that $\tau^2(p) = g + \tau^1(p)$ for all $p \in X$. This proves, using the notation in the statement of the theorem, that $\tau'(p)$ is uniquely of the form $g + \tau\,(p)$ for some $g \in \hat{\mathfrak{g}}$, independent of $p \in X$.

Conversely let $f \in \hat{\mathfrak{g}}$. Since $\lambda_X(x) = \phi^x \circ \tau$ defines a lifting of $d\sigma_X$ so does $\lambda_X{}^f$, given by $\lambda_X{}^f(x) = \lambda_X(x) + \langle f, x \rangle$. That is, (X, ω_X, λ_X) and $(X, \omega_X, \lambda_X{}^f)$ define the structure of a Hamiltonian G-space. If τ_X and $\tau_X{}^f$ are the corresponding orbit coverings (in the sense of Theorem 5.4.1) one has by (5.4.2) $\tau_X = \tau$ and $\tau_X{}^f = \tau^f$ so that 0^f is an orbit and $\tau^f : X \longrightarrow 0^f$ is an orbit covering. QED

For an important family of cases, in looking at the orbit coverings, one obtains all symplectic homogeneous spaces.

Corollary to Theorem 5.5.1. Assume $H^1(\mathfrak{g}, \mathbb{R}) = H^2(\mathfrak{g}, \mathbb{R}) = 0$ (e.g. if $\mathfrak{g}$ is semi-simple). Then the most general G-symplectic homogeneous space (X, ω_X) covers an orbit 0 in $\mathfrak{g}'$. Moreover the covering map τ and hence the orbit are unique.

Proof. Assume (X, ω_X) is a G-symplectic homogeneous space. Since $H^1(\mathfrak{g}, \mathbb{R}) = 0$ one has $\mathfrak{g} = [\mathfrak{g}, \mathfrak{g}]$ so that X is strongly symplectic (see Proposition 4.4.1). On the other hand, since $H^2(\mathfrak{g}, \mathbb{R}) = 0$ then a lifting λ_X to σ_X exists (see Remark 4.5.1). Thus (X, ω_X, λ_X) defines the structure of a Hamiltonian G-space. The result then follows from Theorems 5.4.1 and 5.5.1. QED

Remark 5.5.1. Wang [4] has proved that the most general Kahler homogeneous space for a compact semi-simple Lie group is of the form

G/H where H is the centralizer of a torus in G. The corollary above extends this in several ways. First of all Kahler implies symplectic so the corollary applies. The conclusion of the corollary is that such a space covers an orbit. But since the orbits are simply-connected and the orbits are of the form G/H where H is the centralizer of a torus, one obtains Wang's result with only the assumption of being symplectic. Also one obtains a result without assuming compactness.

5.6. Now let (X, ω_X, λ_X) be a fixed Hamiltonian G-space. We consider the question as to whether $[\omega_X] \in H^2(X, \underline{\underline{R}})$ is integral (so that $\mathcal{L}_c(X, \omega_X)$ is not empty).

For any $p \in X$, let G_p be the isotropic subgroup of G at p so that one has an isomorphism

$$X \cong G/G_p$$

as G-homogeneous spaces.

Now let $f = \tau_X(p) \in \mathfrak{g}'$ and let $O = \tau_X(X) \subseteq \mathfrak{g}'$ so that $O = G \cdot f$ and

$$O \cong G/G_f$$

as G-homogeneous spaces. Since $\tau_X \colon X \longrightarrow O$ is a local diffeomorphism one has

$$(G_f)_o \subseteq G_p \subseteq G_f \tag{5.6.1}$$

where $(G_f)_o$ is the identity component of G_f. Thus in particular

$$(G_p)_o = (G_f)_o \tag{5.6.2}$$

and hence $\mathfrak{g}_f$ is the Lie algebra of both G_f and G_p .

Remark 5.6.1. If we consider the collection of all objects $Y \in \mathrm{Ham}(G)$ such that $\tau_Y Y = 0$ then by Remark 5.1.2 the groups G_p where $p \in \tau_Y^{-1}(f)$ clearly range over all subgroups between G_f and $(G_f)_o$. Of course Y is simply connected if and only if $G_p = (G_f)_o$.

Now by definition of $\mathfrak{g}_f$ clearly f vanishes on $[\mathfrak{g}_f, \mathfrak{g}]$ and hence certainly on $[\mathfrak{g}_f, \mathfrak{g}_f]$. Thus regarding $i\,\mathbb{R}$ as the Lie algebra of the circle group $\mathbb{T}$

$$2\pi i\, f\colon \mathfrak{g}_f \longrightarrow i\,\mathbb{R} \tag{5.6.3}$$

is a homomorphism of Lie algebras.

One obvious question is whether there exists a homomorphism $\Lambda_o : (G_f)_o \longrightarrow \mathbb{T}$, that is a character of $(G_f)_o$, having (5.6.3) as its differential, or even if Λ_o exists, does there exist an extension $\Lambda\colon G_p \longrightarrow \mathbb{T}$ of Λ_o.

If G_p^* is the character group of G_p we let (possibly empty) $G_p^{\#} \subseteq G_p^*$ be the set of all characters $\Lambda\colon G_p \longrightarrow \mathbb{T}$ such that for any $x \in \mathfrak{g}_f$ one has

$$\frac{d}{dt}\Lambda(\exp t\,x)\bigg|_{t=0} = 2\pi i \langle f,x\rangle . \tag{5.6.4}$$

We may take $Y = G/(G_p)_o$ as the simply connected covering space of X where the covering map

$$\tau : Y \longrightarrow X$$

is given by $\tau\big(a\,(G_f)_o\big) = a \cdot p$. Regard Y as an object in $\mathrm{Ham}(G)$ where $\omega_Y = \tau^* \omega_X$ and $\lambda_Y(x) = \lambda_X(x) \circ \tau$ for any $x \in \mathfrak{g}$. We may then identify

$$\Pi = G_p/(G_f)_o \tag{5.6.5}$$

where Π is the fundamental group of X. If $b = a(G_f)_o \in \Pi$ so that $a \in G_p$ then the deck transformation

$$\tau_b : Y \longrightarrow Y \tag{5.6.6}$$

defined by b is given by $\tau_b\big(g(G_f)_o\big) = g\,a\,(G_f)_o$ for $g \in G$. (Recall that G_p normalizes $(G_f)_o$).

By (5.6.5) we may regard the character group Π^* of Π as the subgroup of all characters $\chi \in G_p^*$ such that $\chi | (G_f)_o$ is the trivial character of $(G_f)_o$.

Now clearly $\chi\Lambda \in G_p^\#$ for any $\Lambda \in G_p^\#$ so that Π^* operates on $G_p^\#$ if the latter is not empty.

Proposition 5.6.1. <u>If $G_p^\#$ is not empty then $G_p^\#$ is a Π^*-principal homogeneous space. That is if $\Lambda \in G^\#$ then the map $\Pi^* \longrightarrow G_p^\#$, given by $\chi \longrightarrow \chi\Lambda$, is a bijection</u>.

<u>Proof</u>. This is immediate since if $G_p^\#$ is not empty, it is obviously a coset of Π^* in G_p^*. Indeed this follows since a character on the connected group $(G_f)_o$ is determined by its differential. QED

5.7. Now if $[\omega_X]$ is integral and $\ell = [(L,\alpha)] \in \mathcal{L}_c(X, \omega_X)$ one has by Theorem 4.5.1 the distinguished lifting

$$\sigma_L : G \longrightarrow E(L,\alpha)$$

of σ_X given so that

$$(d\sigma_L)(x) = (\tilde{\delta} \circ \lambda_X (x))$$

for any $x \in \mathfrak{g}$ (see (4.5.3)). We note that if $[(L_i, \alpha_i)] = \ell$ where $i = 1,2$ and

$$\sigma: L_1 \longrightarrow L_2$$

defines an equivalence of line bundles with connection over X then by (4.3.4) one has $(\sigma_X)_*(d\sigma_{L_1})(x) = d\sigma_{L_2}(x)$ so that

$$\sigma\, \sigma_{L_1}(a)\, \sigma^{-1} = \sigma_{L_2}(a) \tag{5.7.1}$$

for any $a \in G$.

Now clearly L_p is stable under $\sigma_L(G_p)$ so that one defines a character $\Lambda^\ell \in G_p^*$ on G_p by putting

$$\Lambda^\ell(a) = \frac{\sigma_L(a)\, u}{u} \tag{5.7.2}$$

for any $u \in L_p^*$. This is obviously independent of u. By (5.7.1) one notes Λ^ℓ does in fact depend only on the class $\ell = [(L, \alpha)]$ and not just on (L, α).

Proposition 5.7.1. *One has* $\Lambda^\ell \in G_p^\#$.

Proof. Let $x \in \mathfrak{g}_f$. Then $(\xi^x)_f = 0$ by Proposition 5.2.1 and (5.3.1) so that $\left(d\sigma_X(x) \right)_p = 0$. Thus the global trajectory γ of $d\sigma_X(x)$ through p is just the point p or $\gamma(t) = p$ for all $t \in \underline{\underline{R}}$. Hence if $s(t)$ is the unique auto-parallel section along γ such that $s(o) = u \in L_p^*$ then $s(t) = u$ for all t.

But if $\tilde{\gamma}(t)$ is the global trajectory of $d\sigma_L(x)$ through u then $\tilde{\gamma}(t) = \sigma_L(\exp - t\, x)\, u$ for any t. However $d\,\sigma_L(x) = \eta\left(\lambda_X(x),\, d\,\sigma_X(x)\right)$ in the notation of (4.2.1) and (4.5.3) and hence by Theorem 2.10.1 one has $\tilde{\gamma}(t) = g(t)\, u$ where

$$g(t) = e^{-2\pi i \int_0^t \left(\lambda_X(x)\right)\left(\gamma(r)\right) dr}$$

But $\left(\lambda_X(x)\right)\left(\gamma(r)\right) = \left(\lambda_X(x)\right)(p) = \lambda_0(x)\,(f) = \langle f,x \rangle$. Hence

$$g(t) = e^{-2\pi i\, t \langle f,x \rangle} \quad \text{or}$$

$$\sigma_L(\exp t\, x)\, u = e^{2\pi i\, t \langle f,x \rangle}\, u$$

for all $t \in \underline{\underline{R}}$ or

$$\Lambda^{\ell}(\exp tx) = e^{2\pi i\, t \langle f,x \rangle}.$$

Hence $(d\,\Lambda^{\ell})(x) = 2\pi i \langle f,x \rangle$ so that $\Lambda^{\ell} \in G_p^{\#}$. QED

Now conversely assume that $G_p^{\#}$ is not empty. Let $\Lambda^{\ell} \in G_p^{\#}$. One constructs a line bundle $L = L_\Lambda$ over X associated to Λ as follows: The group G_p operates freely on $G \times \underline{\underline{C}}$ where $a \cdot (g,z) = \left(g\, a^{-1},\, \Lambda(a)\, z\right)$. One defines L to be the space of G_p orbits in $G \times \underline{\underline{C}}$ where as a manifold L has the quotient structure and one defines $\pi : L \longrightarrow X$ so that $\pi(G_p \cdot (g,z)) = g \cdot p$. The linear structure in $L_p = u^{-1}(p)$ is defined by the linear structure in $\mathbb{C}$. In the usual terminology $L = G \times_{G_p} \mathbb{C}$ is the associated bundle to the principle bundle

$$\begin{array}{ccc} G_p & \longrightarrow & G \\ & & \downarrow \\ & & X \end{array}$$

over X having G as total space and G_p as group and fibre, corresponding to $\Lambda : G_p \longrightarrow \mathrm{Aut}\, \mathbb{C}$. The existence of local smooth sections of the map $G \longrightarrow X$ guarantee that L satisfies the condition of § 1.1.

Now $\mathbb{C}^*$ operates on L so that $c(G_p \cdot (g,z)) = G_p \cdot (g, cz)$ and G operates on L so that $a \cdot (G_p \cdot (g,z)) = G_p \cdot (ag, z)$. Since G commutes with the action of $\mathbb{C}^*$ the action of G induces a homomorphism

$$\sigma_L : G \longrightarrow E(L) .$$

We now observe that L^* has a natural connection form α. Since the action of G and $\mathbb{C}^*$ commute, the product group $H = G \times \mathbb{C}^*$ operates on L and clearly operates transitively on L^* inducing a surjection

$$\nu : H \longrightarrow L^*$$

of H-homogeneous spaces where $\nu(g,z) = G_p \cdot (g,z) = (g,z) \cdot u$ where $u = G_p \cdot (e,1) \in L_p^*$. The group H_u of all elements in H of the form $(a, \Lambda(a)^{-1})$ where $a \in G_p$ is clearly the isotropy subgroup at $u \in G_p \cdot (e, 1) \in L_p^*$.

Now regard $\underline{\underline{C}}$ as the Lie algebra of $\underline{\underline{C}}^*$ (where $\exp z = e^z$) so that the Lie algebra of H is $\underline{h} = \mathfrak{g} \oplus \underline{\underline{C}}$. Any element g in the complex dual to $\underline{h}$ therefore defines a left invariant (Maurer-Cartan) complex 1-form $\delta_g \in \Omega^1(H)$ whose value at $(e,1)$ is g . If $d \in \mathrm{Hom}_{\underline{\underline{C}}}(\underline{\underline{C}},\underline{\underline{C}})$ is such that $\langle d,z \rangle = \frac{z}{2\pi i}$ and $g = (f,d)$ then

$$\delta_g = \left(\delta_f, \frac{1}{2\pi i} \frac{dz}{z}\right) \tag{5.7.3}$$

where δ_f is the left invariant 1-form on G whose value at e is $f = \tau_X p$.

Now from the structure of H_u given above the Lie algebra h_u of H_u is the set of all elements in $\mathfrak{g} \oplus \underline{\underline{C}} = \underline{h}$ of the form $(x, -2\pi i \langle f,x \rangle)$ for all $x \in \mathfrak{g}_f$. But since $g = (f,d)$ it follows that

$$g \mid h_u = 0 \text{ .}$$

Now the differential of ν at the identity of H induces an exact sequence

$$0 \longrightarrow h_u \longrightarrow h \xrightarrow{\nu_*} T_u(L^*) \longrightarrow 0 \text{ .}$$

Hence there exists a unique element g_u in the cotangent space to L^* at u such that $g_u \circ \nu_* = g$.

But now since $G_p \subseteq G_f$ the element $f \in \mathfrak{g}'$ is invariant (relative to Ad') under G_p . Since $\underline{\underline{C}}^*$ is central in H it follows therefore that $g \in \mathrm{Hom}_R(\underline{h}, \underline{\underline{C}})$ in invariant under $\mathrm{Ad}'(H_u)$. Thus g_u is invariant under the action of H_u on the cotangent space at u

and hence there exists a unique 1-form α on L^* which (1) is H-invariant and satisfies $\alpha_u = g_u$. Clearly

$$\nu^*\alpha = \delta_g \tag{5.7.4}$$

so that from the local product structure α is smooth.

Now let $q \in X$ so that $q = a \cdot p$ for some $a \in G$. Then $\nu(a,1) = v \in L_q^*$. But then if $\tau_v : \underline{\underline{C}}^* \longrightarrow L_q^*$ is given by putting $\tau_v(c) = cv$, then $\tau_v = \nu \circ \rho$ where $\rho : \underline{\underline{C}}^* \longrightarrow H$ is given by putting $\rho(c) = (a,c)$. Hence $\tau_v^*\alpha = \rho^*\nu^*\alpha = \rho^*\delta_g$ by (5.7.4). But $\rho^*\delta_g = \frac{1}{2\pi i}\frac{dz}{z}$ by (5.7.3). Thus $\alpha \mid L_q^* = \beta_q$ in the notation of § 1.5. But since also $\underline{\underline{C}}^* \subseteq H$ it follows that α is $\underline{\underline{C}}^*$-invariant and hence (see § 1.5) α defines a connection form on L^*.

Now since the character Λ is $\underline{\underline{T}}$-valued there exists a Hermitian structure N on L such that for $g \in G$, $c \in \underline{\underline{C}}^*$

$$|N|^2 \left(G_p \cdot (g,c)\right) = |N|^2 \left(\nu(g,c)\right) = |c|^2$$

so that $\nu^*\left(\frac{d|N|^2}{|N|^2}\right) = \left(0, \frac{d|z|^2}{|z|^2}\right)$. But since f is real it follows from (5.7.3) and (5.7.4) that

$$\nu^*\left(2\pi i\,(\alpha - \bar{\alpha})\right) = 2\pi i\,(\delta_g - \delta_{\bar{g}}) = \left(0, \frac{d|z|^2}{|z|^2}\right).$$ But from the local product structure ν^* is injective so that

$$\frac{d|N|^2}{|N|^2} = 2\pi i\,(\alpha - \bar{\alpha}).$$

Hence N is an α-invariant Hermitian structure so that

$\ell = (L, \alpha) \in \mathcal{L}_c(X)$. But now $G \leq H$ and since α is H-invariant it is G-invariant. Also N is clearly G-invariant so that

$$\sigma_L : G \longrightarrow E(L, \alpha) \quad .$$

Proposition 5.7.2. <u>Assume</u> $G_p^{\#}$ <u>is not empty</u>. <u>Let</u> $\Lambda \in G_p^{\#}$ <u>and</u> $[(L, \alpha)] = \ell \in \mathcal{L}_c(X)$ <u>be defined as above</u>. <u>Then</u>

$$\omega_X = \text{curv}\ (L, \alpha)$$

<u>so that</u> $\ell \in \mathcal{L}_c(X, \omega_X)$.

<u>Proof</u>. Let $\omega = \text{curv}\ (L, \alpha)$ and let $\rho : G \longrightarrow X$ be the surjection given by putting $\rho(a) = a \cdot p$. Also let as above δ_f be the left invariant 1-form on G whose value at e is f . We assert that

$$\rho^* \omega = d\,\delta_f \quad . \tag{5.7.5}$$

Indeed let the notation be as above so that $\nu: H \longrightarrow L^*$ and $\pi: L \longrightarrow X$. Now let $i : G \longrightarrow H$ be the injection map $a \longrightarrow (a,1)$. Then clearly $\pi \circ \nu \circ i = \rho$. Hence $\rho^*\omega = (i^* \circ \nu^* \circ \pi^*)\omega$. But $\pi^*\omega$ on L^* is $d\alpha$ and hence $\nu^* \pi^* \omega = d\,\delta_g = (d\,\delta_f, 0)$. This proves (5.7.4).

Now if $x \in \mathfrak{g}$ let Γ_x be the (right invariant) vector field on G given so that if $\phi \in C(G)$ and $a \in G$ one has $(\Gamma_x \phi)(a) = \frac{d}{dt} \phi(\exp - tx)a \Big|_{t = 0}$. Then clearly Γ_x and $\sigma_X(x)$ are ρ-related. Thus if $x,y \in \mathfrak{g}$

$$\omega\left(\sigma_X(y)\ ,\ \sigma_X(x)\right) \circ \rho = d\ \delta_f\ \left(\Gamma_y, \Gamma_x\right)$$

$$= \left\langle \left(i\ (\Gamma_y)\ d\ \delta_f\right)\ ,\ \Gamma_x \right\rangle .$$

But $i(\Gamma_y)\ d = -\ d\ i(\Gamma_y) + \theta(\Gamma_y)$. However $\theta(\Gamma_y)\delta_f = 0$ since δ_f is left invariant and Γ_y being right invariant is the infinitesmal generator of a 1-parameter of left translations. Thus $i\ (\Gamma_y)\ d\ \delta_f = -\ d\ \left(i\ (\Gamma_y)\ \delta_f\right) = -\ d \left\langle \delta_f,\ \Gamma_y \right\rangle$. Hence

$$\omega\left(\sigma_X(y),\ \sigma_X(x)\right) \circ \rho = -\ \Gamma_x\ \left\langle \delta_f,\ \Gamma_y \right\rangle .$$

But since $\theta(\Gamma_x)\delta_f = 0$ one has $\Gamma_x \left\langle \delta_f,\ \Gamma_y \right\rangle = \left\langle \delta_f,\ [\Gamma_x,\ \Gamma_y] \right\rangle = \left\langle \delta_f, \Gamma_{[x,y]} \right\rangle$ by (1.3.2). Thus

$$\omega\left(\sigma_X(y),\ \sigma_X(x)\right) \circ \rho = -\left\langle \delta_f,\ \Gamma_{[x,y]} \right\rangle .$$

But $\delta_f = f$ at e and $\Gamma_{[x,y]} = -\ [x,y]$ at e . Thus

$$\omega_p\left(\sigma_X(y),\ \sigma_X(x)\right) = \left\langle f,\ [y,x] \right\rangle$$

$$= (\omega_0)_f\ \left(\sigma_0(y),\ \sigma_0(x)\right)$$

by (5.3.3).

However $\tau_X^*\omega_0 = \omega_X$ and $\sigma_0(x)$ and $\sigma_X(x)$ are τ_X-related so that this expression also equals $(\omega_X)_p\left(\sigma_X(y),\ \sigma_X(x)\right)$. But then $\omega_p = (\omega_X)_p$ since the $\left(\sigma_X(x)\right)_p$, over all $x \in \mathfrak{g}$, span $T_p(X)$. But now ω and ω_X are both G-invariant 2-forms on X . Hence $\omega = \omega_X$. QED

We can prove

Theorem 5.7.1. Let (X, ω_X, λ_X) be any Hamiltonian G-space. Let $p \in X$ and G_p be the isotropy group at p. Then the cohomology class $[\omega_X] \in H^2(X, \underline{R})$ is integral if and only if the set of characters $G_p^{\#}$ (see (5.6.4)) is not empty. Moreover in such a case if Λ^ℓ is the character on G_p (see (5.7.2)) defined by an element $\ell \in \mathcal{L}_c(X, \omega_X)$ then $\Lambda^\ell \in G_p^{\#}$ and the map

$$\mathcal{L}_c(X, \omega_X) \longrightarrow G_p^{\#} \tag{5.7.6}$$

given by $\ell \longrightarrow \Lambda^\ell$ is a bijection. We recall that $\mathcal{L}_c(X, \omega_X)$ is the set of equivalences of all line bundles with connection over X having ω_X as curvature. Furthermore (5.7.6) is a Π^*-map where Π^* is the character group of the fundamental group of X, recalling that $\mathcal{L}_c(X, \omega_X)$ and $G_p^{\#}$ are principal Π^*-homogeneous spaces by Theorem 2.5.1 and Proposition 5.6.1.

Proof. If there exists $\Lambda \in G_p^{\#}$ then by Proposition 5.7.2 the constructed (L, α) defines an element ℓ in $\mathcal{L}_c(X, \omega_X)$. By Proposition 2.1.1 this implies that $[\omega_X]$ is integral. Conversely if $[\omega_X]$ is integral there exists $\ell = [(L, \alpha)] \in \mathcal{L}_c(X, \omega_X)$. Hence by Proposition 5.7.1 the character Λ^ℓ lies in $G_p^{\#}$ so that $G_p^{\#}$ is not empty.

Now assume $[\omega_X]$ is integral. Since by Theorem 2.5.1 and Proposition 5.6.1 both $\mathcal{L}_c(X, \omega_X)$ and $G_p^{\#}$ are principal Π^*-homogeneous spaces we have only to prove that 5.7.6 is a Π^* map or that

$$\Lambda^{\chi\ell} = \chi\Lambda^\ell . \tag{5.7.7}$$

for any $\chi \in \Pi^*$ and any $[(L, \alpha)] = \ell \in \mathcal{L}_c(X, \omega_X)$.

We use the notation of § 2.4, 2.5 and § 5.6. Here $X^1 = G/(G_f)_o$ is a simply connected covering space to X where the covering map β of § 2.4 is τ of § 5.6. Also $\omega_{X_1} = \omega^1 = \beta^*\omega_X$. Obviously $[\omega^1]$ is integral since $[\omega_X]$ is integral. The homomorphism $\sigma\colon \Pi \longrightarrow D(X^1, \omega^1)$ (see (2.4.1)) is given by $\sigma(b) = \tau_{b^{-1}}$ (see (5.6.6)). Now let (L^1, α^1) be the (unique with curvature ω^1) line bundle with connection over X^1 as in § 2.4 and let

$$\nu\colon \Pi \longrightarrow E(L^1, \alpha^1) \tag{5.7.8}$$

be a lifting of σ (see Proposition 2.4.1). As in § 2.4 the quotient space $L^1/\nu(\Pi) = L$ is a line bundle over X and has a connection form α such that $\tilde{\beta}^*\alpha = \alpha^1$ where $\tilde{\beta}\colon L^1 \longrightarrow L$ is the quotient map. One has $\ell_\nu = [(L, \alpha)] \in \mathcal{L}_c(X, \omega_X)$ in the notation (2.4.3). We wish now to compute the character $\Lambda^{\ell_\nu} \in G_p^*$ in terms of ν.

Now G operates on L^1 and L relative to the homomorphisms

$$\sigma_{L_1} : G \longrightarrow E(L^1, \alpha^1) \quad \text{and} \quad \sigma_L : G \longrightarrow E(L, \alpha)$$

given by Theorem 5.1.1 (since (X, ω_X, λ_X) and $(X^1, \omega_{X^1}, \lambda_{X^1})$ are Hamiltonian G-spaces). We assert that $\tilde{\beta} : L^1 \longrightarrow L$ is a G-map. By Remark 4.4.1 we have only to see that $\tilde{\beta}$ is a $\mathfrak{g}$-map. But since $x \in \mathfrak{g}$ and $d\,\sigma_L = \tilde{\delta}_L \circ \lambda_X$, $d\,\sigma_{L^1} = \tilde{\delta}_{L^1} \circ \lambda_{X^1}$ and $\lambda_{X^1}(x) = \lambda_X(x) \circ \beta$ for $x \in \mathfrak{g}$, we have only to show that $\tilde{\delta}_L(\phi)$ is $\tilde{\beta}$-related to $\tilde{\delta}_{L^1}(\phi \circ \beta)$ for any $\phi \in R(X)$. But ξ_ϕ is β-related to $\xi_{\phi \circ \beta}$

since β is a local symplectic isomorphism. But since $(\beta)^* \alpha = \alpha^1$, β_* carries horizontal vectors to horizontal vectors so that ξ_ϕ and $\xi_{\phi \circ \beta}$ are $\tilde{\beta}$-related by the commutative diagram (2.4.2). But now if $v \in (L^1)^*$ then $\left(\eta(\phi \circ \beta)\right)_v$ is the tangent vector to the curve $t \longrightarrow e^{-2\pi i\, t\phi\left(\beta(\pi^1 v)\right)} v$ at $t = 0$ and $\eta(\phi)_{\tilde{\beta}v}$ is the tangent vector to the curve $t \longrightarrow e^{-2\pi i\, t\phi\, (\pi\tilde{\beta}v)} \tilde{\beta}v$ at $t = 0$. But then $\eta(\phi)$ are $\eta(\phi \circ \beta)$ are $\tilde{\beta}$-related since $\tilde{\beta}$ is linear on $L^1{}_v$ and $\pi\tilde{\beta} = \beta\pi^1$ by (2.4.2). Thus $\delta_L(\phi) = \eta(\phi) + \tilde{\xi}_\phi$ is $\tilde{\beta}$-related to $\delta_{L^1}(\phi \circ \beta) = \eta(\phi \circ \beta) + \tilde{\xi}_{\phi \circ \beta}$ so that $\tilde{\beta}$ is a G-map.

Now let $u \in L_p$ so that by (5.7.2)

$$\Lambda^{\ell}\nu(a) = \frac{\sigma_L(a)\, u}{u} \tag{5.7.9}$$

for any $a \in G_p$. Now if p^1 is the coset $(G_f)_o$ in X^1 then $\beta(p^1) = p$. Let $v \in L^1{}_{p^1}$ be such that $\tilde{\beta}v = u$ and let $b = a(G_f)_o \in \Pi$. But now

$$\tilde{\beta}\left(\nu(b^{-1})\, v\right) = \tilde{\beta}(v) = u \tag{5.7.10}$$

since $\tilde{\beta} \circ \nu(b^{-1}) = \tilde{\beta}$ by definition of $\tilde{\beta}$. However $\nu(b^{-1})\, v \in L^1{}_{\sigma(b^{-1})p^1}$. But $\sigma(b^{-1})\, p^1 = \tau_b\, p^1$ and $\tau_b\, p^1$ is the coset $a(G_f)_o$ in X^1. On the other hand

$$\tilde{\beta}\left(\sigma_{L^1}(a)\, v\right) = \sigma_L(a)\, u \tag{5.7.11}$$

since $\tilde{\beta}$ is a G-map. But $\sigma_{L^1}(a)\, v \in L_{\sigma_{L^1}(a)\,\cdot\, p^1}$ and $\sigma_{X^1}(a)\, p^1$

is also the coset $a(G_f)_o$. Thus $\frac{\sigma_{L^1}(a)\, v}{\nu(b^{-1})v}$ is well defined and since $\tilde{\beta}$ is linear on $L^1_{\sigma_{X^1}(a)} \cdot p^1$ one has by (5.7.9), (5.7.10) and (5.7.11) that

$$\Lambda^{\ell_\nu}(a) = \frac{\sigma_{L^1}(a)\, v}{\nu(b^{-1})v} \quad . \tag{5.7.12}$$

But now if $\chi \in \Pi^*$ then $\chi\, \ell_\nu = \ell_{\chi\nu}$ by (2.5.6).
But $(\chi\nu)(b^{-1}) = \chi(b^{-1})\, \nu(b^{-1})$ by Proposition 1.13.1.
But also $\chi(b) = \chi(a)$ where Π^* is regarded as a subgroup of G_p^* .
Thus substituting $\chi\nu$ for ν in (5.7.12) one has

$$\Lambda^{\chi\ell_\nu}(a) = \chi(a)\, \Lambda^{\ell_\nu}(a) = (\chi\Lambda)^{\ell_\nu}(a) \quad .$$

Since (L^1, α^1), v and σ_{L^1} are independent of the lifting ν , and since $\ell = \ell_\nu$ is an arbitrary element of $\mathcal{L}_c(X, \omega_X)$ one has

$$\Lambda^{\chi\ell} = \chi\Lambda^{\ell}$$

for all $\chi \in \Pi^*$, $\ell \in \mathcal{L}_c(X, \omega_X)$. QED

Corollary 1 to Theorem 5.7.1. <u>Let G be any simply connected Lie group and let $\mathfrak{g}$ be its Lie algebra. Let $0 \subseteq \mathfrak{g}'$ be a G-orbit so that $(0, \omega_0, \lambda_0)$ is a Hamiltonian G-space</u> (see Theorem 5.3.1). <u>Let $f \in 0$ and let G_f be the isotropy group at f . Then the cohomology class $[\omega_0] \in H^2(0, \mathbb{R})$ is integral if and only if there exists a character Λ on G_f whose differential is $2\pi i\, f \mid \mathfrak{g}_f$. Here $\mathfrak{g}_f$ is the Lie algebra of G_f . Moreover in such a case if $G_f^{\#}$ is the set of all such characters then $G_f^{\#}$ and $\mathcal{L}_c(0, \omega_0)$, the set of all</u>

line bundles with connection over O having ω_O as curvature, are both principal homogeneous space for Π^*, the character group of the fundamental group Π of O. Also to each $[(L, \alpha)] = \ell \in \mathcal{L}_c(O, \omega_O)$ one has an element $\Lambda^\ell \in G_f^\#$ defined as in (5.7.2) by restricting the lifting $\sigma_L(G_f)$ of the action of G_f in O to the line L_f. Finally the correspondence $\mathcal{L}_c(O, \omega_O) \longrightarrow G_f^\#$ given by $\ell \longrightarrow \Lambda^\ell$ is a Π^*-map and hence is a bijection.

Proof. Follows from Theorem 5.7.1 by taking X to be an orbit in $\mathfrak{g}'$. See Theorem 5.3.1. QED

REFERENCES

1. R. Godement, Théorie des faisceaux, Hermann, Paris.

2. R. Palais, Lie theory of Transformation Groups, Memoirs of the Amer. Math Soc. 22, (1957).

3. I. E. Segal, Quantization of Non-Linear Systems, Journal of Math. Physics I (1960) pp. 468-488.

4. H. C. Wang, Closed manifolds with homogeneous structures, Amer. Jour. Math. 76 (1954), 1-32.

5. A. Weil, Variétés kahlériennes, Hermann, Paris.

Lectures in Modern Analysis and Applications I

CONTENTS

MODERN METHODS AND NEW RESULTS IN COMPLEX ANALYSIS

Professor KENNETH M. HOFFMAN, Massachusetts Institute of Technology

Bounded Analytic Functions in the Unit Disc 1

A discussion of the compactification of the unit disc, which is induced by the algebra of bounded analytic functions, especially Carleson's work on the corona theorem and the speaker's work on analytic subsets of the compactification.

Professor HUGO ROSSI, Brandeis University

Strongly Pseudoconvex Domains 10

Some of the most exciting work in several complex variables done in the past ten years centers around the solution of Levi's problem: to show that a pseudoconvex domain is holomorphically convex. Pseudoconvexity is a differential condition on the nature of the boundary; the latter implies the existence of many holomorphic functions. The key to the solution is the theorem of finite dimensionality of the cohomology groups of a coherent sheaf, proven by Grauert, Kohn, Hörmander. From this one can fully describe the analytic structure of a strongly pseudoconvex domain, and this gives rise to a method for studying isolated singularities of analytic spaces.

BANACH ALGEBRAS AND APPLICATIONS

Professor CHARLES E. RICKART, Yale University

A function algebra, which does not contain all continuous functions, may exhibit certain properties reminiscent of analycity. An example is the local maximum modulus principle proved by Hugo Rossi. This, along with various results, suggests the beginnings of an abstract analytic function theory. At this stage, the program is to obtain analogues of certain results from several complex variables.

TOPOLOGICAL LINEAR SPACES AND APPLICATIONS

Professor LARS HÖRMANDER, Institute for Advanced Study

For the Cauchy problem with data on a hyperplane there exists a unique solution for srbitrary data if and only if the equation is hyperbolic in the sense of Garding. When the hyperplane is characteristic there is no longer uniqueness, but we characterize the equations having a solution for arbitrary Cauchy data. This class contains all parabolic equations.

Professor F. TREVES, Purdue University

Local Cauchy problems for systems of linear PDEs with analytic coefficients, with data on noncharacteristic hypersurfaces, have always unique solutions. But these in general need not be distributions, they are ultradistributions. A simple proof of this fact is possible, based on general results about abstract differential equations (also valid for nonlinear ones), suitably adapted blowing up of small domains and ladders of functional (Banach) spaces. This allows a detailed description of the situation, including of the symbols of "fundamental solutions", and reveals the links with the problem of solvability in more classical sense.

GEOMETRIC AND QUALITATIVE ASPECTS OF ANALYSIS

Professor MICHAEL F. ATIYAH, Institute for Advanced Study and Oxford University

Certain spaces of operators in Hilbert space have interesting connections with the theory of vector bundles in algebraic topology. This lies at the root of recent work on the topological aspects of elliptic differential equations.

Professors CLIFFORD J. EARLE and JAMES EELLS, Cornell University

Two elliptic operators play a fundamental role in the transcendental deformation theory of Riemann surfaces: Beltrami's equation (of conformal geometry) and the tension equation (of the theory of harmonic maps). Their properties are used to construct certain fibre bundles belonging to the theory; in particular, Teichmüller space appears as a universal principle bundle.

Professor STEPHEN SMALE, University of California, Berkeley

When can a differential equation be perturbed and still retain many of its qualitative properties? This leads to questions of structural stability and Ω-stability which are investigated.

*This lecture was presented by Professor James Eells.

Lectures in Modern Analysis and Applications II

CONTENTS

ANALYSIS AND REPRESENTATION THEORY

Professor HARISH-CHANDRA, Institute for Advanced Study

Some Applications of the Schwartz Space of a Semisimple Lie Group.......... 1

Let G be a semisimple Lie group. The analogue for G of the usual Schwartz space (on R^n) plays an important role in the harmonic analysis on G.

Professor RICHARD V. KADISON, University of Pennsylvania

Some Analytic Methods in the Theory of Operator Algebras.................. 8

Recent theorems on groups of automorphisms of C^* and von Neumann algebras will be discussed together with their relation to modern physical theory.

MODERN ANALYSIS AND NEW PHYSICAL THEORIES

Professor IRVING E. SEGAL, Massachusetts Institute of Technology

The Mathematical Theory of Quantum Fields..................................30

A contemporary mathematical view of quantum field theory, with answers for such questions as: what is a quantum field?, and illustrations from quantum electrodynamics. The formulation of the S-matrix as the "time-ordered exponential of the interaction Hamiltonian" will be discussed.

Professor JAMES GLIMM, Courant Institute of Mathematical Sciences, New York University

Mathematical Problems in the Foundations of Quantum Field Theory...........58

Mathematical problems presented by quantum field theory. The occurence of infinities and the features of the physics responsible for their presence. Approximations to the correct laws of physics which remove the infinities. Removal of all approximations to obtain as a limit a quantum field Φ with a Φ^4 selfinteraction in two dimensional space time.

Recent developments in the statistical mechanics of infinite systems will be discussed. Some insight into the meaning of the Gibbs phase rule is obtained.

INTEGRATION IN FUNCTION SPACES AND APPLICATIONS

Basic definitions and concepts in the theory of integration over infinite dimensional vector spaces will be described. Newtonian potential theory on Euclidean n-space will be extended to potential theory on an infinite dimensional Hilbert space as an application and motivation. It will be shown how Wiener space arises from these considerations.

Lecture Notes in Mathematics

Bisher erschienen/Already published

Vol. 1: J. Wermer, Seminar über Funktionen-Algebren. IV, 30 Seiten. 1964. DM 3,80 / $ 1.10

Vol. 2: A. Borel, Cohomologie des espaces localement compacts d'après. J. Leray. IV, 93 pages. 1964. DM 9,– / $ 2.60

Vol. 3: J. F. Adams, Stable Homotopy Theory. Third edition. IV, 78 pages. 1969. DM 8,– / $ 2.20

Vol. 4: M. Arkowitz and C. R. Curjel, Groups of Homotopy Classes. 2nd. revised edition. IV, 36 pages. 1967. DM 4,80 / $ 1.40

Vol. 5: J.-P. Serre, Cohomologie Galoisienne. Troisième édition. VIII, 214 pages. 1965. DM 18,– / $ 5.00

Vol. 6: H. Hermes, Term Logic with Choise Operator. III, 55 pages. 1970. DM 6,– / $ 1.70

Vol. 7: Ph. Tondeur, Introduction to Lie Groups and Transformation Groups. Second edition. VIII, 176 pages. 1969. DM 14,– / $ 3.80

Vol. 8: G. Fichera, Linear Elliptic Differential Systems and Eigenvalue Problems. IV, 176 pages. 1965. DM 13,50 / $ 3.80

Vol. 9: P. L. Ivănescu, Pseudo-Boolean Programming and Applications. IV, 50 pages. 1965. DM 4,80 / $ 1.40

Vol. 10: H. Lüneburg, Die Suzukigruppen und ihre Geometrien. VI, 111 Seiten. 1965. DM 8,– / $ 2.20

Vol. 11: J.-P. Serre, Algèbre Locale. Multiplicités. Rédigé par P. Gabriel. Seconde édition. VIII, 192 pages. 1965. DM 12,– / $ 3.30

Vol. 12: A. Dold, Halbexakte Homotopiefunktoren. II, 157 Seiten. 1966. DM 12,– / $ 3.30

Vol. 13: E. Thomas, Seminar on Fiber Spaces. IV, 45 pages. 1966. DM 4,80 / $ 1.40

Vol. 14: H. Werner, Vorlesung über Approximationstheorie. IV, 184 Seiten und 12 Seiten Anhang. 1966. DM 14,– / $ 3.90

Vol. 15: F. Oort, Commutative Group Schemes. VI, 133 pages. 1966. DM 9,80 / $ 2.70

Vol. 16: J. Pfanzagl and W. Pierlo, Compact Systems of Sets. IV, 48 pages. 1966. DM 5,80 / $ 1.60

Vol. 17: C. Müller, Spherical Harmonics. IV, 46 pages. 1966. DM 5,– / $ 1.40

Vol. 18: H.-B. Brinkmann und D. Puppe, Kategorien und Funktoren. XII, 107 Seiten, 1966. DM 8,– / $ 2.20

Vol. 19: G. Stolzenberg, Volumes, Limits and Extensions of Analytic Varieties. IV, 45 pages. 1966. DM 5,40 / $ 1.50

Vol. 20: R. Hartshorne, Residues and Duality. VIII, 423 pages. 1966. DM 20,– / $ 5.50

Vol. 21: Seminar on Complex Multiplication. By A. Borel, S. Chowla, C. S. Herz, K. Iwasawa, J.-P. Serre. IV, 102 pages. 1966. DM 8,– / $ 2.20

Vol. 22: H. Bauer, Harmonische Räume und ihre Potentialtheorie. IV, 175 Seiten. 1966. DM 14,– / $ 3.90

Vol. 23: P. L. Ivănescu and S. Rudeanu, Pseudo-Boolean Methods for Bivalent Programming. 120 pages. 1966. DM 10,– / $ 2.80

Vol. 24: J. Lambek, Completions of Categories. IV, 69 pages. 1966. DM 6,80 / $ 1.90

Vol. 25: R. Narasimhan, Introduction to the Theory of Analytic Spaces. IV, 143 pages. 1966. DM 10,– / $ 2.80

Vol. 26: P.-A. Meyer, Processus de Markov. IV, 190 pages. 1967. DM 15,– / $ 4.20

Vol. 27: H. P. Künzi und S. T. Tan, Lineare Optimierung großer Systeme. VI, 121 Seiten. 1966. DM 12,– / $ 3.30

Vol. 28: P. E. Conner and E. E. Floyd, The Relation of Cobordism to K-Theories. VIII, 112 pages. 1966. DM 9,80 / $ 2.70

Vol. 29: K. Chandrasekharan, Einführung in die Analytische Zahlentheorie. VI, 199 Seiten. 1966. DM 16,80 / $ 4.70

Vol. 30: A. Frölicher and W. Bucher, Calculus in Vector Spaces without Norm. X, 146 pages. 1966. DM 12,– / $ 3.30

Vol. 31: Symposium on Probability Methods in Analysis. Chairman. D. A. Kappos. IV, 329 pages. 1967. DM 20,– / $ 5.50

Vol. 32: M. André, Méthode Simpliciale en Algèbre Homologique et Algèbre Commutative. IV, 122 pages. 1967. DM 12,– / $ 3.30

Vol. 33: G. I. Targonski, Seminar on Functional Operators and Equations. IV, 110 pages. 1967. DM 10,– / $ 2.80

Vol. 34: G. E. Bredon, Equivariant Cohomology Theories. VI, 64 pages. 1967. DM 6,80 / $ 1.90

Vol. 35: N. P. Bhatia and G. P. Szegö, Dynamical Systems. Stability Theory and Applications. VI, 416 pages. 1967. DM 24,– / $ 6.60

Vol. 36: A. Borel, Topics in the Homology Theory of Fibre Bundles. VI, 95 pages. 1967. DM 9,– / $ 2.50

Vol. 37: R. B. Jensen, Modelle der Mengenlehre. X, 176 Seiten. 1967. DM 14,– / $ 3.90

Vol. 38: R. Berger, R. Kiehl, E. Kunz und H.-J. Nastold, Differentialrechnung in der analytischen Geometrie IV, 134 Seiten. 1967 DM 12,– / $ 3.30

Vol. 39: Séminaire de Probabilités I. II, 189 pages. 1967. DM 14,– / $ 3.90

Vol. 40: J. Tits, Tabellen zu den einfachen Lie Gruppen und ihren Darstellungen. VI, 53 Seiten. 1967. DM 6.80 / $ 1.90

Vol. 41: A. Grothendieck, Local Cohomology. VI, 106 pages. 1967. DM 10,– / $ 2.80

Vol. 42: J. F. Berglund and K. H. Hofmann, Compact Semitopological Semigroups and Weakly Almost Periodic Functions. VI, 160 pages. 1967. DM 12,– / $ 3.30

Vol. 43: D. G. Quillen, Homotopical Algebra. VI, 157 pages. 1967. DM 14,– / $ 3.90

Vol. 44: K. Urbanik, Lectures on Prediction Theory. IV, 50 pages. 1967. DM 5,80 / $ 1.60

Vol. 45: A. Wilansky, Topics in Functional Analysis. VI, 102 pages. 1967. DM 9,60 / $ 2.70

Vol. 46: P. E. Conner, Seminar on Periodic Maps. IV, 116 pages. 1967. DM 10,60 / $ 3.00

Vol. 47: Reports of the Midwest Category Seminar I. IV, 181 pages. 1967. DM 14,80 / $ 4.10

Vol. 48: G. de Rham, S. Maumary et M. A. Kervaire, Torsion et Type Simple d'Homotopie. IV, 101 pages. 1967. DM 9,60 / $ 2.70

Vol. 49: C. Faith, Lectures on Injective Modules and Quotient Rings. XVI, 140 pages. 1967. DM 12,80 / $ 3.60

Vol. 50: L. Zalcman, Analytic Capacity and Rational Approximation. VI, 155 pages. 1968. DM 13.20 / $ 3.70

Vol. 51: Séminaire de Probabilités II. IV, 199 pages. 1968. DM 14,– / $ 3.90

Vol. 52: D. J. Simms, Lie Groups and Quantum Mechanics. IV, 90 pages. 1968. DM 8,– / $ 2.20

Vol. 53: J. Cerf, Sur les difféomorphismes de la sphère de dimension trois $(\Gamma_4 = 0)$. XII, 133 pages. 1968. DM 12,– / $ 3.30

Vol. 54: G. Shimura, Automorphic Functions and Number Theory. VI, 69 pages. 1968. DM 8,– / $ 2.20

Vol. 55: D. Gromoll, W. Klingenberg und W. Meyer, Riemannsche Geometrie im Großen. VI, 287 Seiten. 1968. DM 20,– / $ 5.50

Vol. 56: K. Floret und J. Wloka, Einführung in die Theorie der lokalkonvexen Räume. VIII, 194 Seiten. 1968. DM 16,– / $ 4.40

Vol. 57: F. Hirzebruch und K. H. Mayer, O (n)-Mannigfaltigkeiten, exotische Sphären und Singularitäten. IV, 132 Seiten. 1968. DM 10,80 / $ 3.00

Vol. 58: Kuramochi Boundaries of Riemann Surfaces. IV, 102 pages. 1968. DM 9,60 / $ 2.70

Vol. 59: K. Jänich, Differenzierbare G-Mannigfaltigkeiten. VI, 89 Seiten. 1968. DM 8,– / $ 2.20

Vol. 60: Seminar on Differential Equations and Dynamical Systems. Edited by G. S. Jones. VI, 106 pages. 1968. DM 9,60 / $ 2.70

Vol. 61: Reports of the Midwest Category Seminar II. IV, 91 pages. 1968. DM 9,60 / $ 2.70

Vol. 62: Harish-Chandra, Automorphic Forms on Semisimple Lie Groups X, 138 pages. 1968. DM 14,– / $ 3.90

Vol. 63: F. Albrecht, Topics in Control Theory. IV, 65 pages. 1968. DM 6,80 / $ 1.90

Vol. 64: H. Berens, Interpolationsmethoden zur Behandlung von Approximationsprozessen auf Banachräumen. VI, 90 Seiten. 1968. DM 8,– / $ 2.20

Vol. 65: D. Kölzow, Differentiation von Maßen. XII, 102 Seiten. 1968. DM 8,– / $ 2.20

Vol. 66: D. Ferus, Totale Absolutkrümmung in Differentialgeometrie und -topologie. VI, 85 Seiten. 1968. DM 8,– / $ 2.20

Vol. 67: F. Kamber and P. Tondeur, Flat Manifolds. IV, 53 pages. 1968. DM 5,80 / $ 1.60

Vol. 68: N. Boboc et P. Mustată, Espaces harmoniques associés aux opérateurs différentiels linéaires du second ordre de type elliptique. VI, 95 pages. 1968. DM 8,60 / $ 2.40

Vol. 69: Seminar über Potentialtheorie. Herausgegeben von H. Bauer. VI, 180 Seiten. 1968. DM 14,80 / $ 4.10

Vol. 70: Proceedings of the Summer School in Logic. Edited by M. H. Löb. IV, 331 pages. 1968. DM 20,– / $ 5.50

Vol. 71: Séminaire Pierre Lelong (Analyse), Année 1967 – 1968. VI, 19 pages. 1968. DM 14,– / $ 3.90

Bitte wenden / Continued

Beschaffenheit der Manuskripte

Die Manuskripte werden photomechanisch vervielfältigt; sie müssen daher in sauberer Schreibmaschinenschrift geschrieben sein. Handschriftliche Formeln bitte nur mit schwarzer Tusche eintragen. Notwendige Korrekturen sind bei dem bereits geschriebenen Text entweder durch Überkleben des alten Textes vorzunehmen oder aber müssen die zu korrigierenden Stellen mit weißem Korrekturlack abgedeckt werden. Falls das Manuskript oder Teile desselben neu geschrieben werden müssen, ist der Verlag bereit, dem Autor bei Erscheinen seines Bandes einen angemessenen Betrag zu zahlen. Die Autoren erhalten 75 Freiexemplare.

Zur Erreichung eines möglichst optimalen Reproduktionsergebnisses ist es erwünscht, daß bei der vorgesehenen Verkleinerung der Manuskripte der Text auf einer Seite in der Breite möglichst 18 cm und in der Höhe 26,5 cm nicht überschreitet. Entsprechende Satzspiegelvordrucke werden vom Verlag gern auf Anforderung zur Verfügung gestellt.

Manuskripte, in englischer, deutscher oder französischer Sprache abgefaßt, nimmt Prof. Dr. A. Dold, Mathematisches Institut der Universität Heidelberg, Tiergartenstraße oder Prof. Dr. B. Eckmann, Eidgenössische Technische Hochschule, Zürich, entgegen.

Cette série a pour but de donner des informations rapides, de niveau élevé, sur des développements récents en mathématiques, aussi bien dans la recherche que dans l'enseignement supérieur. On prévoit de publier

1. des versions préliminaires de travaux originaux et de monographies
2. des cours spéciaux portant sur un domaine nouveau ou sur des aspects nouveaux de domaines classiques
3. des rapports de séminaires
4. des conférences faites à des congrès ou à des colloquiums

En outre il est prévu de publier dans cette série, si la demande le justifie, des rapports de séminaires et des cours multicopiés ailleurs mais déjà épuisés.

Dans l'intérêt d'une diffusion rapide, les contributions auront souvent un caractère provisoire; le cas échéant, les démonstrations ne seront données que dans les grandes lignes. Les travaux présentés pourront également paraître ailleurs. Une réserve suffisante d'exemplaires sera toujours disponible. En permettant aux personnes intéressées d'être informées plus rapidement, les éditeurs Springer espèrent, par cette série de »prépublications«, rendre d'appréciables services aux instituts de mathématiques. Les annonces dans les revues spécialisées, les inscriptions aux catalogues et les copyrights rendront plus facile aux bibliothèques la tâche de réunir une documentation complète.

Présentation des manuscrits

Les manuscrits, étant reproduits par procédé photomécanique, doivent être soigneusement dactylographiés. Il est recommandé d'écrire à l'encre de Chine noire les formules non dactylographiées. Les corrections nécessaires doivent être effectuées soit par collage du nouveau texte sur l'ancien soit en recouvrant les endroits à corriger par du verni correcteur blanc.

S'il s'avère nécessaire d'écrire de nouveau le manuscrit, soit complètement, soit en partie, la maison d'édition se déclare prête à verser à l'auteur, lors de la parution du volume, le montant des frais correspondants. Les auteurs recoivent 75 exemplaires gratuits.

Pour obtenir une reproduction optimale il est désirable que le texte dactylographié sur une page ne dépasse pas 26,5 cm en hauteur et 18 cm en largeur. Sur demande la maison d'édition met à la disposition des auteurs du papier spécialement préparé.

Les manuscrits en anglais, allemand ou francais peuvent être adressés au Prof. Dr. A. Dold, Mathematisches Institut der Universität Heidelberg, Tiergartenstraße ou au Prof. Dr. B. Eckmann, Eidgenössische Technische Hochschule, Zürich.

ISBN 3-540-05284-4
ISBN 0-387-05284-4